High Above

Chris Forrester
Editor

High Above

The untold story of Astra, Europe's leading satellite company

Springer

Editor
Chris Forrester
Broadgate Publications
37 The Towers
Lower Mortlake Road
Richmond, Surrey
TW9 2JR, UK

Second Printing, Originally published by Broadgate Publications, UK, 2010

ISBN 978-3-642-12008-4 e-ISBN 978-3-642-12009-1
DOI 10.1007/978-3-642-12009-1
Springer Heidelberg Dordrecht London New York

Library of Congress Control Number: 2011934054

Cover design: eStudio Calamar S.L

Printed on acid-free paper

Springer is part of Springer Science+Business Media (www.springer.com)

Our Digital Age

Chris Forrester

Frequently it is suggested that the 'golden age' of television was during the period 1950-1960. It is true that television almost ruined Hollywood's fortunes during this period. But if this was the authentic golden age, then it was an age of black and white, somewhat limited creativity, poor reception, lack of competition (except in the United States) and – by and large – public service broadcasting.

In 1950 television was at its most mature in the United States. Some three million sets were sold in the USA during the first six months of that year. Four networks competed for viewer loyalty (NBC, CBS, ABC and DuMont), although NBC and CBS dominated. The DuMont network went out of business in 1956 and ABC struggled through TV's first ten years before finding its feet in the 1960s.

In war-ravaged Europe progress was significantly slower. France established Radiodiffusion-Télévision Française (RTF) in 1945, immediately after the liberation but did not initiate full weekday television until 1947. Its millionth TV receiver was not sold until 1958, despite the development in 1956 of its Sequential Coleur à Memoire (SECAM) colour system.

Germany had managed to maintain a pattern of propaganda TV broadcasting to public places until 1944, but it wasn't until 1949 and the establishment of the Federal Republic that 'modern' television could even begin to blossom. Control of TV was ceded to the various regional Länder, although the whole system was hamstrung by the notorious Proporz method, whereby political parties allocated senior TV posts in direct proportion to their parliamentary powers; the programming chief coming from Party A, news chief from Party B, etc.

An Astra satellite, covered in highly reflective gold foil but without its all-important solar panels, is shown in a test chamber, at the Astrium plant in Toulouse, France

ARD, West Germany's first public network (although supported by advertising), was joined by ZDF in 1962, again with commercial time sold in specific time slots, but not within programmes.

Television broadcasting in Britain recommenced in 1946, under the firm hand of John Reith. With a new Conservative government elected in 1951 approval was given for a second, commercial service to be established with a franchised system of regional operators, licensed by the (then) Independent Television Authority. If sales of television sets are any guide then the coronation of Queen Elizabeth II in 1953 marked the beginning of Britain's television age.

Japan's television engineers had pioneered many developments pre-war, but television was initiated by public broadcaster NHK on February 1, 1953, and a commercial service (Nippon TV) on August, 28. Services were limited to Tokyo, and not surprisingly receivers were limited – just 866 at year-end. Imported from the USA, a set would typically cost 250,000 Yen, equivalent to the annual salary of a middle-class white-collar worker. Another royal event, the wedding of Crown Prince Akihito to Michiko Shoda (April 10, 1959), was again the catalyst in selling an estimated 2 million sets.

SES Astra is a partner of YahSat in YahLive, which has capacity on YahSat 1A

However, if we take 1950 as a generic 'starting point' for modern television broadcasting, then we talk about a kind of prehistoric stage of the medium – in which it remained for the best part of three decades. The real youth, as we saw, of broadcasting were the 1980s; the time when commercial television started on a large scale (and not only in Luxembourg) and, in this youth, was getting younger in terms of programming.

Astra appeared on the scene at exactly this time. Astra has not only been around since then but it was instrumental for the dramatic development that we have witnessed since then.

This is the story we want to tell in this book. While it would be simple to categorise television's development over this period, the progress from analogue to digital to HDTV represents a technological quantum leap that depended on a number of crucial factors coming together. It's also worth remembering the development and maturity of the analogue system was itself a key element in this progress. Without multi-channel analogue TV, the mass-market penetration of television sets and general economic prosperity we would not have the necessary base ingredients to make the great leap forward into digital, into HDTV, 3D-television, and the prospects of Ultra High Definition now in sight.

The story of satellite is therefore more than a story of technology. It is more than a story of rockets, spacecraft, solar panels, transponders, footprints, decoders and dishes. It is the story of the birth of a whole array of new industries and sectors, from hardware manufacturing to programme production, and from single network domination to numerous niche channels. It has meant the creation of hundreds of thousands of jobs in these sectors, of millions of hours of programming and billions of fresh market capitalizations at international stock exchanges.

And it is also a reflection of politics, 20th Century history, the Cold War and the collapse of Communism, the creation of Europe, economic growth and wealth, big brands, the Internet age and of technological convergence. It is a period that has seen the emergence of television as the dominant information and entertainment medium. Or, to sum it up differently: a story of a century. And a story of success.

'High Above' tells the story of Astra, and how some extremely brave pioneers managed to overcome technical, political and commercial obstacles and in the process become the world's leading satellite operator. And a catalyst for dramatic change in the lives of all of us who watch TV.

'High Above' tells the story of Astra, and how some extremly brave pioneers managed to overcome technical, political and commercial obstacles and become the world`s leading satellite operator.

01

TV that 'Comes Alive'

01
TV that 'Comes Alive'

Chris Forrester

Digital. Convergence. Super-Highway. MPEG. MHEG. Multichannel. The Web, HDTV, 3D, Broadband, Ultra-HDTV. Despite the television industry being packed full of jargon and buzz words, we are an industry that has to constantly look to the future, while remembering the lessons of the past. Pick whichever buzz word you like, it really doesn't matter because our digital future is here now and key to all these developments is satellite. All we have to do is open our eyes and recognise the signs. People consume content and their appetite grows larger day by day. And satellite delivers that content.

How Canal+ sees its offerings

1985

85/01/21
Ronald Reagan sworn in for second term

C. Forrester (ed.), *High Above: The Untold story of Astra, Europe's leading satellite company*,
DOI 10.1007/978-3-642-12009-1_1, © Springer-Verlag Berlin Heidelberg 2011

Apple´s colourful view of the future: Apple iPad

To help compile this overview of the satellite and broadcasting industry today we have taken the liberty to dip back 10 or 20 years or so into our archives and see what the forecasters were then predicting. Let me quote the truly prophetic Ray Hammond. In his benchmark 'The On-Line Handbook' (1984) he says, "The linking of computers around the world is going to have far-reaching effects, and the spread of knowledge, the interchange of ideas and the dissemination of information are going to produce a revolution in our society."

Hammond's words showed how dangerous it is to attempt to be a latter-day Nostradamus. Hammond was absolutely accurate about the wired future, just a few years off on his estimate of how long it would take for us to get hooked. To his credit, in his follow-up volume ('Digital Business', 1996), Hammond admits that the network revolution had taken longer to achieve critical mass.

Across the Atlantic there are even more pundits and prophets. Pulitzer Prize-winning writer Joel Brinkley, in his 'Defining Vision' (1997), speaking of the digital TV revolution, said: "These new machines are wondrous indeed because, unlike the earlier Japanese models, these are digital televisions. A TV that receives its signal digitally is no longer just a dumb box passively displaying pictures and sound. Digital televisions, properly equipped, can be powerful, interactive computers, hardly different from desktop PCs. With those capabilities, suddenly television comes alive."

Which is all very well, but as broadcasting crosses the threshold into the Century's second decade (and it seems like only yesterday that we were all welcoming in Year 2000) it is worth remembering that most things that have 'digital' attached to their names take longer to bed in with the mass market. Read Abe Peled, a true television visionary and chairman of technology company NDS. He spoke in 1999: "Look further ahead, look past the current Sky Digital system, which is already the best in the world. Look to even greater fragmentation of the audience. Take the USA, perhaps the most radical example of this, where over the past 10 years some 50% of the primetime audience has disappeared from the major networks. But it's not that they're sitting in front of their PCs, but they are watching one of the dozens of other cable networks. It hasn't changed that much, simply fragmented in sub-sets of an audience. It makes it more difficult for advertisers to reach that mass audience, but in many respects, digital allows more targeted advertising, even down to post code advertising. BMWs for Manhattan, and Range Rovers for the country."

Peled, as usual, is right. But even 10 years after his forecast, the fact is that the world is only just taking its first tentative steps into targeted TV advertising into the home. Despite more than 10 million addressable satellite set-top boxes in the UK, and tens of millions more across Europe and further afield, the advertising community, as well as brand owners, needed to catch up with the broadcast technology. Satellite will again be key in reaching these valued viewers with uniquely targeted advertising – just as it today delivers millions of hours of thematic programming.

But Peled was right in one prediction during the same interview. He spoke of the

85/03/01
Société Européenne des Satellites (SES) founded in Luxembourg

85/02/19
British TV soap EastEnders debuts

85/03/11
Mikhail Gorbatchev appointed head of Soviet Union

The normally accepted truism of broadcasting, that "content is king", is only partially true. More accurate in a digital, multi-channel, multi-platform age might be "the trusted brand is king".

tumbling fall in the cost of chipsets, which Peled said already cost "nothing and next year will be next to nothing." He said that hard-drive storage costs are also in freefall, and this makes successful digital technology possible. "Cheap silicon chips will make digital set-top boxes even more affordable," says Peled.

However, most developed countries of the world are already migrating to digital transmission, and the technology has caught on enthusiastically with most emerging nations. They too want to re-allocate analogue bandwidth for alternative or allegedly more lucrative uses, and one way or another the world's broadcasters are embracing digital transmission. And while all viewers may not yet be enjoying commonplace HDTV transmissions, critics say that there seems to be an inexorable slide towards whatever the Hollywood studios want. If they say HDTV, then it will be HDTV for all. If they say 3D then it will be 3D for all – at least on some channels. It will just take a little longer to achieve mass-market penetration for these technologies. Inexpensive chip-sets, lower-cost flat screen televisions, free or inexpensive set-top boxes, and not to forget Dolby multi-channel surround sound (said by more than a few experts to have been responsible for selling more large-screen TVs than any other technological advance since colour), the cost of entertainment, at least as far as the hardware is concerned, is falling. But convergence, from the World Wide Web and cheap telephony is also influencing every element of the broadcasting mix.

Take these words from Microsoft's Jim Beveridge (in September 1999): "[The industry is] going to be using more fibre, more ADSL, more cable, more satellite and more wireless. In my view we are going to be using a mixture of broadcast standards and IP-based standards. I see broadcast and IP protocol-based standards coming together. If you are thinking of what the consumer is going to use, then what the consumer is going to have is a number of gateways to getting programming to him. Some will be IP-based and some will be broadcast-based, and I actually think the methods that will work will be when the two base standards come together."

He added: "What people will do is take elements of that which is offered and build into their own networks what they believe is worth including."

There's the clue. Unlimited satellite bandwidth, web-streaming, cellular... Ten years ago we began to see quickly-emerging Palm-type and Windows-CE personal devices emerge that also stored video, but as 2010 unfolds, it is now near impossible to escape the video-based appeal of cellular and the promise of 4G transmission. Indeed, listen to the creative people at Nokia, Ericsson, Motorola and Microsoft as they talk about 'next generation' cellular telephony with harmonised standards now agreed upon. We are already seeing cell phones that can surf the Web, transmit video, cope with TV and recognise human speech for ease of operation. Far more is promised, and in some markets

Reflection from above: IFA technology trade fair on September 4, 2009 in Berlin, Germany

the availability of satellite-delivered content to handheld (and vehicle-based) content is very much a reality.

4G and Hybrid Developments

Players like Microsoft are not alone. There are many cellular operators who also supply video (and frequently radio as well). Broadband suppliers take the same view. Terrestrial broadcasters – thanks to digital transmission – are also increasingly sophisticated in the services they can offer. It is the same with satellite, as with France`s TNTSAT, which is serving and complementing digital terrestrial installations to reach full geographical coverage where conventional broadcast towers are extremely cost-inefficient. The music industry seeks to tap into video, while 'next generation' radio broadcasters also see video as being a potential money spinner. A recent study from consultants Arthur D. Little suggests that the strongest combination for a converged service will end up being a hybrid set-top box that supplies TV from satellite as well as broadband Internet from terrestrial sources.

And these are not so-called 'blue sky' developments. All of the above products exist. One challenge still to be overcome by all these mobile-based technologies, no matter how they operate, are the twin – and usually contradictory – demands of most users. On the one hand users want small cellular devices, yet at the same time manufacturers want more functionality and the moment you build in a 'TV' screen into a handheld device you have an inherent drift towards bulk and high battery power demand.

85/09/24
SES signs contract with RCA for Astra 1A

85/07/13
Live Aid concert

85/09/01
Wreck of Titanic is found

These problems are being solved, but there is little doubt that consumers are demanding ever more sophistication in their handheld devices. And this extends to all members of the family. Ten years ago the prospect of pre-teenagers having their own sophisticated video-enabled 'smart' cellphones was unheard of. Not today!

Technology Success – or Failure

Not all technological introductions have been success stories. It is worth remembering that Apple's iPod is a very recent device (it made its debut only in 2001). There were earlier MP3 players, but they simply failed to catch on. Even the might of Microsoft wasn't enough to ensure that Zune was successful. Despite the hype and support from one of the best-known products on the face of the planet, Zune never achieved popular adoption. Similarly, Yahoo has been around for some 15 years, AOL a little longer. Google, patented in 2001, has only emerged in the past 10 years. YouTube was created only in 2005. The same argument applies to a slew of equally recent creations, not least MySpace, Facebook, and 2009's hot property Twitter. These are the successes. And each and every one of these services can be accessed on computers, on cell phones and increasingly on TV sets. Indeed, there's a huge effort being made by the world's middleware and 'widget' vendors to deliver these – perhaps transient – applications into the living room.

But the fact is that for each technological success there are usually 100 failures. Our list kicks off with Laser Discs, BetaMax (for home use), HD-DVD and even AOL as examples of powerfully-backed products or services that have struggled. Indeed, some might argue that Microsoft's Vista was a similar business failure, but so are products like Psion, Vonage, Joost, Babelgum, Friends Reunited and a dozen other sexy products that fly high for a little while and then crash to Earth. Some struggle on, helped by rich backers but the tough economic reality is that today's consumer is an extremely fickle user.

And of course, behind the scenes, our business is very much focussed on technology. But facing the market, it is usually a question of branding. Take Disney. Think family entertainment and you think of Disney – perhaps it's RTL in Germany or the BBC in the UK but in the USA it's Disney – and when viewers want to experience family entertainment they need to be able to find a Disney-branded product. There are Disney-branded cellular phones for youngsters with amazing functionality. The Disney executives want their brand and content everywhere: In the movie theatre, on TV of course, as well as in packaged media (DVD and Blu-ray). But there are also times when they are on the radio, and in the Theme Park, and on theatre stages around the world, and even with a hockey team playing on a Saturday night.

These are all viable-share businesses. They will be the first choice for building that primary brand value. Disney, as a name, as a brand, is just the starting-off point. However, all of these separate platforms must be seamlessly integrated to present a kind of content continuum. Disney is surely the

Japanese kids love their Disney phones

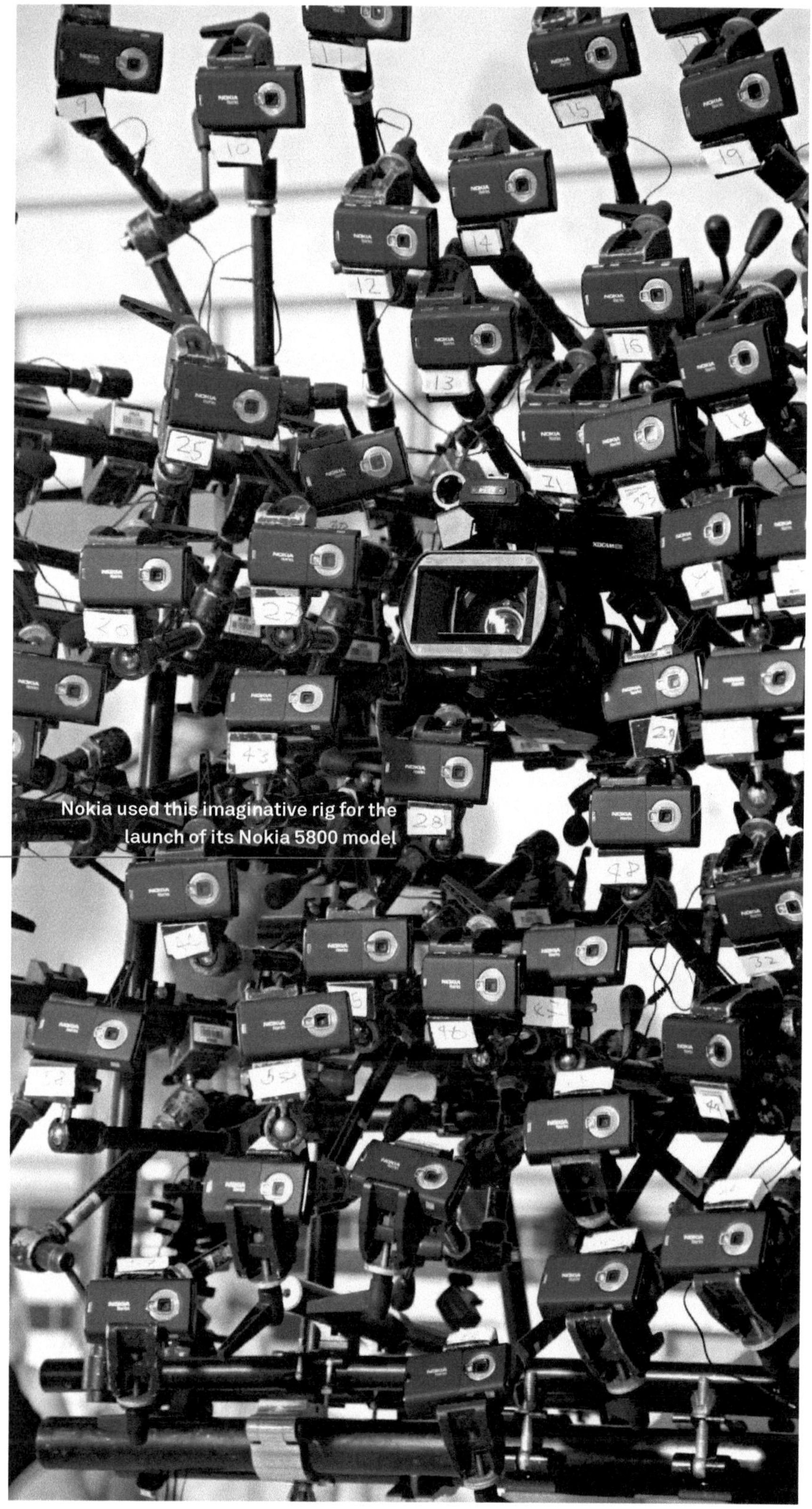

Nokia used this imaginative rig for the launch of its Nokia 5800 model

But the fact is that for each technology success there are usually 100 failures.

85/10/18
First Nintendo Game Boy released in USA

85/11/20
Microsoft releases Windows 1.0

perfect example of brand extension, not content with animation or natural history or even theme parks. Their brand can now be found just everywhere. In cellular-hungry Japan the company has more than 38 million users who have paid for the little mouse's cellular-based content. There's barely a child (or an adult) on the planet who doesn't understand what Disney is all about. Progress was made in movies, of course, but it's television that has delivered a billion eyes to the Disney logo.

Is Content Still King?

The normally accepted truism of broadcasting, that "content is king", is only partially true. More accurate in a digital, multi-channel, multi-platform age might be "the trusted brand is king" as with Disney. But the blur between content and technology, or content and utility, is now so fuzzy that it is difficult to differentiate what is what. Some industry experts have suggested that only established brands have any chance of succeeding in this new world. Perhaps. But look at the three most recognised brands that have recently emerged, none of which are in the 'content' business that we would normally recognise: Amazon, eBay and Google. All three represent utility, and utility has become content in that environment – and that has built their brand value.

We all thought that "content is king" meant an ever-present CNN and the might of appealing programming. A more realistic view is that these are now less relevant and that customers in an interactive, technology-based market – which is what every future platform increasingly offers – cannot tell the difference between CNN, Reuters, Associated Press or Agence France Presse (the video news agencies where CNN's news images frequently come from) or a brand like Yahoo and its news supply. One way to achieve relevance is to supply better entertainment, better features and better technology than anyone else, or to package those utilities, features or technologies better than anyone else. This is what a packaging and portal play is all about: creating platform-specific material via the usual types of media structures.

What is increasingly clear is that broadcasters now recognise that the show itself is key. It is the programme itself that viewers want to tap into. Whether it is Big Brother on YouTube, or a Pop Idol-type show that's streamed on the Web, the broadcaster can be near-invisible. Tony Cohen, president of RTL-owned Fremantle Media, says: "The viewer not only watches TV; he wants to consume it beyond, in a game or even in merchandising." Cohen says that the relationship between brands and producers has evolved enormously through the new media. "Until today, this relationship existed through TV channels. The advertisers need a more emotional link with their target group, and the producers are ideally placed to generate it. In a certain way, the channels no longer have the monopoly on that relationship."

Adam Singer, a well-regarded UK media pundit and at one time chairman of Flextech (now part of Virgin Media), the UK's largest digital broadcaster after BSkyB, argues that just about everything becomes a sub-set of the Internet and broadband. "The Internet's high-speed capacity is capable of carrying [everything]. If that is a true statement then you just pull down from the pipe the information you want and you utilise it on the device that's most relevant to you. This is already done and exists on the telephone. The telephone line is the pipe, and if it happens to be a Fax signal you pull it down to a Fax machine, if it's a paging message it goes to a pager or if it's voice to a telephone.

What we want to show is the pivotal role that satellite played in the whole development of media, its explosion and convergence.

It seems to me the same thing happens [in entertainment]. You have your info-brand and you pull down the bit you want onto the appropriate technology. The one thing you don't want to get hung up on is the 'lean-forward, lean-back' thing-in-the-corner conversation. Because there will be times when you want to be told a story, and times when you want to interact, and times when you want to know more." Singer adds: "You will choose the appropriate device for the time and place. You can already see this on the MP3 players like iPods. What is an MP3 unit, but a portable hard drive. It provides all sorts of opportunities for mainstream broadcasters."

Neither does Singer see there being a "winner takes all" technology emerging. Singer`s message for the infrastructure business is: "It's interesting that there should be any concept of a single winner. The concept doesn't exist with music, where there are umpteen ways of receiving music. You don't often hear the argument that radio was the 'winner' over the concert hall. They are both forms of distribution and dissemination. So, it strikes me that it is all to do with the best way to receive information at any given moment. [...] As soon a you start trying to select a winner, you are lost. You can be certain that is not going to be the winner."
So we will not declare satellite a winner here. This is not the purpose of this book. What we want to show is the pivotal role that satellite played in the whole development of media, its explosion and convergence. Satellite has also been the absolute catalyst in creating enterprises like the set-top box industry, or Conditional Access businesses, or the motivation behind the TV manufacturing sector. Programming rights trading could never have grown to the €10 billion business it is today without satellite's region-wide coverage, syndication opportunities and core viewership – and their need for content. Satellite has also played its part in building truly global brands such as Disney, Discovery, Cartoon Network, CNN, MTV and their many other TV rivals.

Chinese film fans watch 'Avatar' in 3D, January 7, 2010

85/12/27
Terrorists kill 18 at Rome and Vienna airports

02

Location, Location, Location

02
Location, Location, Location

Geoff Bains

In television broadcasting, height is everything. For engineers charged with bringing TV signals to the maximum number of homes, over the widest area, the key is to broadcast from a great height, so the transmissions can 'see' over hills, into valleys and beyond local obstacles. Since the early days of terrestrial television, that meant mounting the transmitters at the top of tall buildings or on even taller masts.

Although TV masts got higher and higher – up to 650 metres in Warsaw – and the transmitters more powerful, it was never enough.

The rise of reliable and affordable satellite technology has provided the answer. Satellites can transmit television signals from space, in orbits around the Earth, where they are far higher than any mast can be built.

Astra's fleet of satellites orbits nearly 36,000 kilometres above the Earth so their signals can reach homes in the deepest valleys or nestled among high-rise buildings, with ease.

It can be difficult to picture just how high 36,000 kilometres really is. It's about 110,000 Eiffel Towers piled up on top of each other – but even that doesn't really bring the distance home. Instead, consider that you were able to drive your car vertically, straight up into the air. Cruising at a steady 100 km/hour (without rest breaks) you would officially become an astronaut after about an hour and a half. After three and a half hours you would overtake the International Space Station and two hours later, you might spot the Hubble Space Telescope. But it would take 15 days of non-stop driving before you could park alongside a geostationary satellite.

1986

86/01/01
Pierre Meyrat joins as SES' first Director General

86/01/01
Spain and Portugal join EU

C. Forrester (ed.), *High Above: The Untold story of Astra, Europe's leading satellite company*, DOI 10.1007/978-3-642-12009-1_2, © Springer-Verlag Berlin Heidelberg 2011

Different television satellites are placed at different points around the geostationary orbital path at 36,000 kilometres above the earth.

At such a height, TV satellites can spread their signal over a far wider area than any Earth-based transmitter. At 36,000 kilometres, the horizon is a long way away and the satellite can 'see' about 40% of the whole surface of the Earth, so orbiting satellites can easily transmit to every country in Europe, and beyond, from a single position in space.

In 1957, Sputnik 1, the very first man-made satellite, was equipped with a simple radio transmitter for reception back on Earth – just like today's TV satellites. However, today's TV satellites have two big advantages over that first craft – their orbit and their transmission power.

Communications satellites take a very special path around the globe, called a geostationary orbit. Like any orbit, this involves the craft circling the world at a high altitude and a fast speed. The higher the orbit, the longer it takes the satellite to complete one full circle around the Earth, and in the geostationary orbit at 36,000 kilometres above the

86/01
The State of Luxembourg guarantees LuxFr 3.6 billion to the SES project

86/01/20
Plans announced to build Channel Tunnel

86/01/28
Space Shuttle 'Challenger' explodes after takeoff

86/02/03
Pixar Animation Studios opened

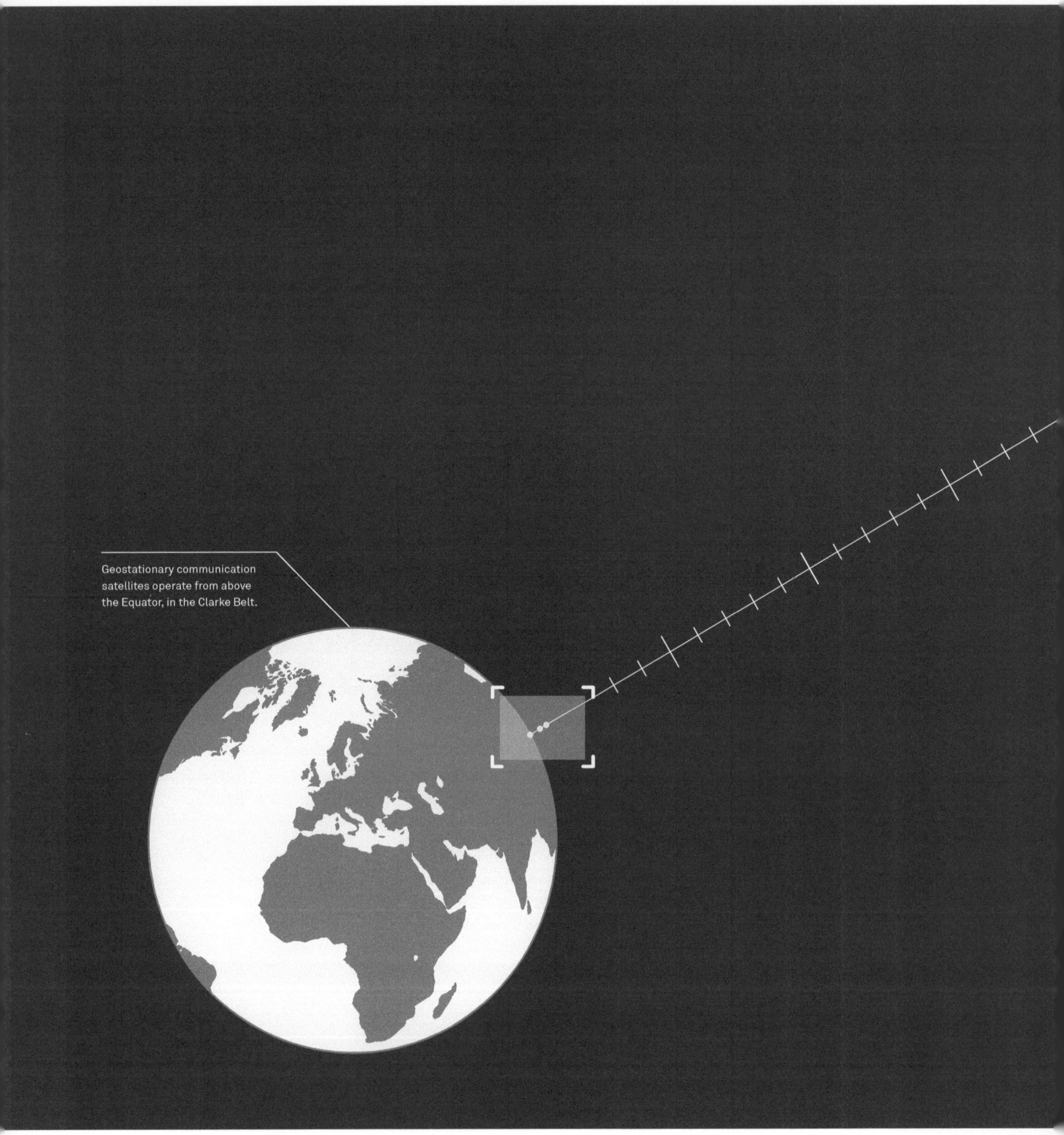
Geostationary communication
satellites operate from above
the Equator, in the Clarke Belt.

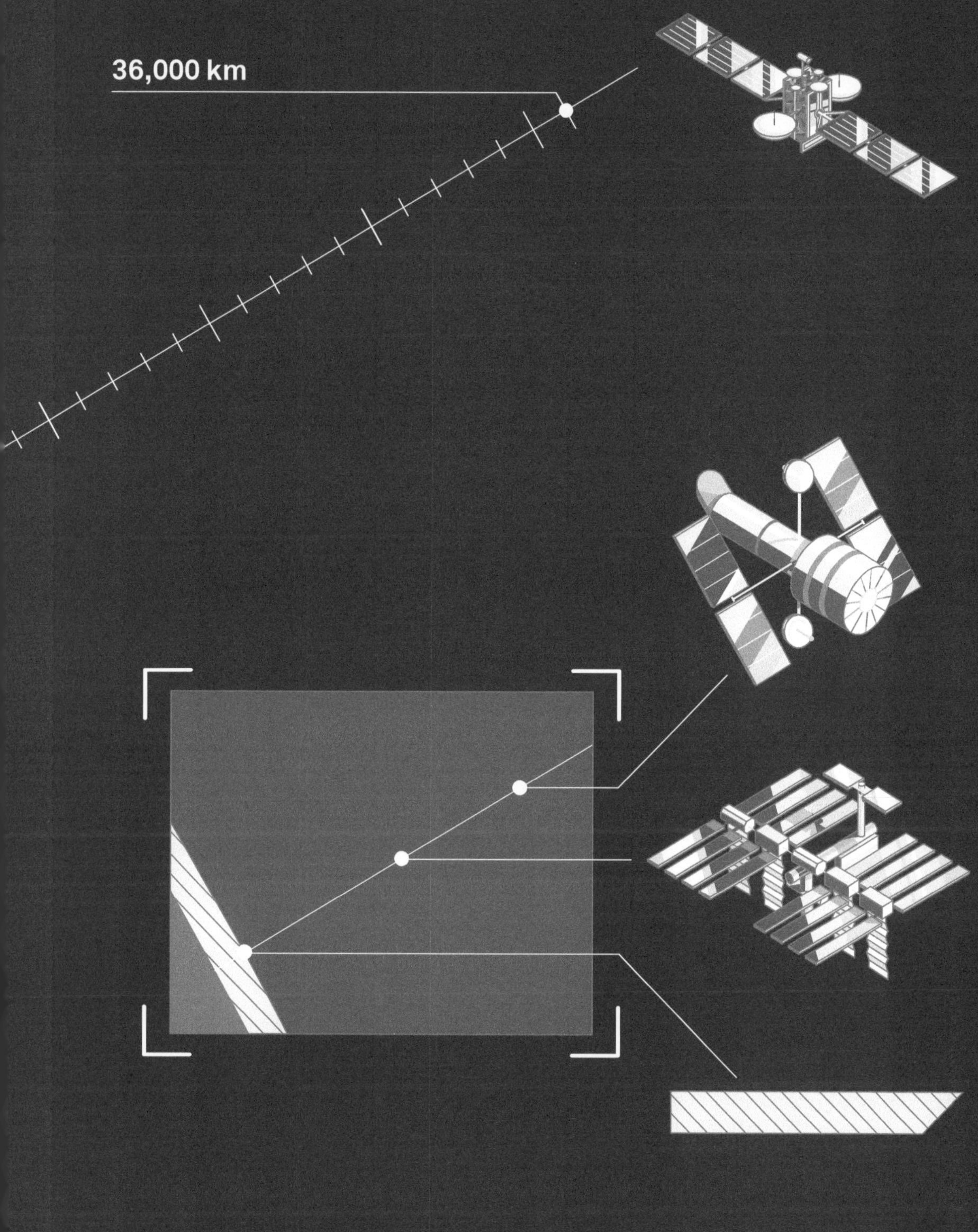

Astra's fleet is firmly in the Geostationary Orbit (GEO) at about 35,863 kilometres high, more than 100 times higher than the International Space Station.

The Hubble Space Telescope operates well above any trace of atmosphere, at 569 km above the Earth's surface. It orbits the Earth every 97 minutes moving at a speed of about 8 km per second.

The International Space Station operates between 319 km and 347 km above the Earth.

The Earth's atmosphere is generally said to be thinning out at the Troposphere, about 17 km up. The Stratosphere extends to about 50 km.

Earth, the satellite takes 24 hours to circle the globe. Of course, this is the same time that it takes the Earth to rotate once on its own axis – the length of one day.

Since the TV satellites are in an orbit directly above the Earth's equator, and they travel towards the east (the same direction that the Earth is turning), the satellites match the movement of the Earth's surface turning beneath. They seem to appear stationary, hanging in the sky, to anyone on the rotating globe beneath.

Different television satellites are placed at different points around the geostationary orbital path. They are all moving at high speed around the Earth, but spread out in a line so they seem fixed at different points in the sky where they can best access different parts of the Earth (Astra's main orbital positions are above equatorial Africa). The entire geostationary orbit is about 265,000 kilometres long, so there is room for plenty of satellites – at the moment, there are about 235 satellites transmitting TV and radio signals from the geostationary orbit.

The use of the geostationary orbit is key to satellite TV, as it means that an antenna on the ground can be pointed at the right point in the sky to receive the required satellite signals, and fixed in place, where it will remain aimed correctly, receiving the broadcast signals as though the transmitter really were on at the top of a 36,000 kilometres high transmission mast.

Although the TV satellites may appear to be at a standstill in the sky, in reality they are far from motionless. A geostationary satellite flies around the planet at a speed of over 11,000 km/h and the gravitational influence of the Moon and the Sun, as well as the imperfect symmetry of the Earth beneath, contribute to slight irregularities in its path. So from the Earth it seems that a satellite drifts from the correct position.

To keep it on-station, small rocket jets on the satellite fire tiny bursts of gas to push the satellite back into position. To conserve the satellite's fuel and prolong its life, the position is not corrected continuously but the craft is allowed to drift within an imaginary box in space, 70 kilometres along each side, which although huge by terrestrial standards, is small enough to avoid the need for the receiving dishes to be constantly re-aligned.

So, operating a television satellite is far from a launch-and-forget exercise, and Astra's fleet of satellites has huge resources devoted to keeping the craft in the right positions so that the millions of dishes aimed at them continue to receive their TV signals. Maintaining one satellite in the right orbital position is challenge enough, but Astra's fleet includes several orbital positions with several satellites together. From the ground, these appear to occupy the same point in the sky but in fact they are kept well apart, dancing around one another in that imaginary box in the sky, like a magical celestial ballet, but where there's zero contact!

This station-keeping is all under the direction of a ground control station, monitoring the satellite's position and transmitting commands to switch on and off the on-board propulsion systems. Astra's control centre is at SES' headquarters at the Château de Betzdorf in Luxembourg, where the engineers study the satellites' movements and issue the commands to nudge each craft back to its correct place.

These station-keeping commands are sent to the satellites using the array of huge dishes alongside the control centre in Betzdorf, and these same dishes also carry the majority of the television signals up to the satellites – for this is really 'all' that a TV satellite does; it takes in signals transmitted from the ground and simultaneously re-broadcasts those signals back to the Earth.

The uplink transmission from Betzdorf to the satellite is at one frequency and the downlink transmission from the satellite to the millions of dishes across Europe is at another frequency, so they don't interfere with one another. In other respects, the signals are identical. What goes up, comes down.

In effect, a satellite is like a huge TV mirror in space, reflecting back to Earth what is transmitted up from the uplink centre, and this is why this system is sometimes described as a 'bent pipe' – the TV signals are turned around by the satellite and sent back to Earth. Of course, the technology to turn around the television signals is not as simple as a mirror, or even a bent pipe, and complex electronics called transponders handle the reception of the uplink signal, its conversion from one frequency to another, and its re-transmission as the downlink.

All geostationary satellites contain several transponders so that they can handle many signals at once. Each signal has a different uplink frequency and a different downlink frequency so a satellite TV receiver in the home can choose between the signals by selecting the frequency received – just like the dial on a simple radio tuner selects a frequency from the broadcast band to 'tune in' to a station.

Satellite TV signals are at a much higher frequency than terrestrial TV or radio transmissions, and among the highest frequencies found in any electronic equipment in the home. They are high-frequency and thus avoid interference with terrestrial transmissions. These signals pass through the atmosphere on their long journey from the satellite at the speed of light and with extremely little degradation.

Each satellite may weigh several tonnes and be the size of a small van, with huge solar panel 'wings' that stretch out many metres on each side.

Each transponder in the satellite broadcasts its signal not just at one particular frequency but over a small range of frequencies, the bandwidth. The wider the bandwidth, the greater the amount of information that can be carried on the signal, so it's preferable to broadcast signals with a wide bandwidth. However, this means that fewer different signals can be 'fitted' into the overall frequency range available.

What's more, the broadcast signal from one transponder is separated from the signals of other transponders by a frequency 'no man's land' to ensure that the signals do not overlap and interfere with one another. This further reduces the number of signals that can be broadcast.

The clever solution is to make use of the available frequency range twice. Signals are broadcast with one of two possible 'polarities' – the direction in which the signals are vibrating. A signal transmitted with horizontal polarity can be distinguished from another with vertical polarity, and so they can use a similar broadcast frequency without interfering with one another.

Satellites stagger the frequencies of each polarity to ensure that there is absolutely no interference, but the use of dual-polarity broadcasting doubles the number of transponders that can successfully broadcast from each satellite. The overall transmission frequency range has been extended as well, and the signals from Astra satellites today occupy more than twice the frequency spectrum than the first Astra satellites could use.

Indeed, Astra satellites were the first in Europe to make use of some parts of today's satellite TV frequencies and such was the importance of receiving channels broadcasting on Astra, that satellite receiver specifications were universally altered to accommodate the changes.

Although we speak of TV satellites in geostationary orbit appearing motionless in the sky, none of the Astra satellites can actually be seen from Earth with the naked eye. Each satellite may weigh several tonnes and be the size of small van, with huge solar panel 'wings' that stretch out many metres on each side, but at nearly 36,000 kilometres from the Earth they are visible only with a telescope.

86/08/19
Picasso's Weeping Woman recovered in Melbourne

86/09/07
Desmond Tutu is first black bishop in South Africa

However, the satellites' transmissions are clearly 'visible' to our TV antennas. It is the second big difference from the earliest spacecraft that enables the Astra satellites to send TV pictures to more than 100 million homes on the Earth's surface below. Unlike Sputnik 1, which carried a single 1W transmitter to broadcast its simple identifying 'beep' to the world, the latest Astra television satellites use transmitters of up to 150W for each transponder, as well as carefully designed reflectors to concentrate all of that power towards a particular area of the Earth for which the transmissions are intended.

Nevertheless, even these transmissions need amplification. The strength of the signal reaching a dish is truly tiny, barely detectable without the modern electronics that make up a home satellite TV reception system. It's something like using a pocket torch on the Statue of Liberty in New York to signal to a colleague on the Eiffel Tower in Paris.

The concave reflector of a satellite dish collects the radio signals from the satellite and focuses them onto the electronics package on the end of the dish arm. This is the LNB, or low noise block-down converter, and the great unsung hero of everyday satellite TV reception. An LNB is a complex and state-of-the-art part of the satellite TV reception equipment, vital to the successful reception of TV signals from space and capable of operating correctly unattended for years in all types of weather, yet is rarely given a moment's thought by the television viewers inside, in their warm living rooms or dens.

The main task of the LNB is to act as an aerial for the incoming signals, and to convert the radio signal to an electrical one. It also has to amplify the electrical signal and change it to a lower frequency so it can be sent down a cable to the satellite receiver box inside the home.

A modern LNB provides amplification of as much as 1,000,000 times, but introduces almost no spurious noise into the television signal. It's also responsible for selecting which of the two polarities of signal – horizontal or vertical – and what part of the extended overall frequency spectrum are to be received and passed on to the receiver. It responds to commands from the receiver to make these selections, without the intervention, or even knowledge, of the user inside watching television from satellite. There's another fact that people frequently forget, and it concerns the absolute reliability of a satellite's signal. In Astra's case they have a record of 99.9998% reliability for their broadcast clients. In other words any signal problems would equate to theoretical outage of just a few minutes per channel but spread over 25 years of operation. Such reliability represents an impressive achievement.

Today, LNBs are small, efficient, and cheap, but it wasn't always so. And huge strides in the development of better and more capable LNBs have been made in the past 21 years, not least for reception of signals from the numerous, and most popular, Astra satellites. It's not for nothing that the most familiar type of these devices, used today on dishes in millions of homes across Europe, is commonly named for the leading protagonist in its development, as an Astra LNB. Twin and quad-LNBs are now commonplace and enable viewers to receive signals from multiple satellites.

The use of the geostationary orbit is key to satellite TV, as it means that an antenna on the ground can be pointed at the right point in the sky to receive the required satellite signals.

86/09/08
Oprah Winfrey show debuts nationwide in US

86/09/15
'LA Law' debuts on NBC

In Astra's case they have a record of 99.9998% reliability for their broadcast clients. In other words any signal problems would equate to a theoretical outage of just a few minutes per channel but spread over 25 years.

One of Astra's more modern satellites, Astra 1L, being integrated into its rocket at Kourou, May 2007

6/10/09
News Corp buys Metromedia. Launches Fox TV

86/11/22
Mike Tyson wins first major title in Las Vegas

86/12/14
Dick Rutan and Jeana Yeager fly Voyager around globe

Ground Control to Major Tom (except there's no Major Tom)

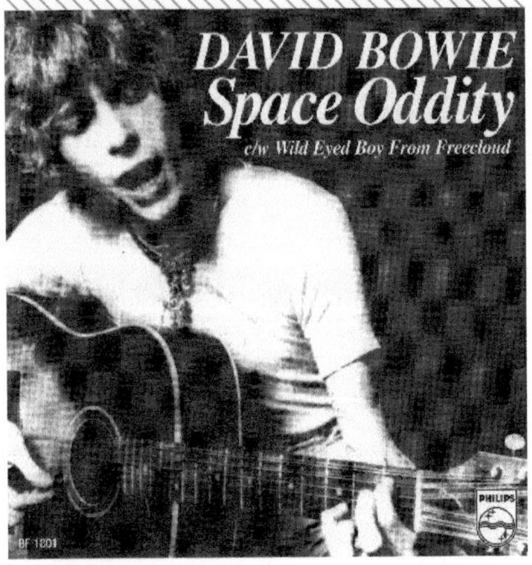

We all know the tune, debuted by David Bowie in his 1969 album 'Space Oddity', when our brave astronaut was told to "take your protein pills and put your helmet on". Great fun, but Astra's orbiting satellites are marvels of complexity and sophistication. Despite often being referred to as 'bent pipe' devices (that simply collect, amplify, and return the signal to Earth) the truth is that they are somewhat beyond the reach of humans other than through their on-board telemetry and control systems.

Geostationary communication satellites orbit at 36,000 kilometres above the Earth's surface and must orbit at a very precise speed. Too fast, or too slow, and they risk escaping their designated orbit. All 'Geo' communication satellites maintain an orbital position and speed that is constant to the Earth's surface. This way a satellite dish always 'sees' the craft at the same position in the sky.

However, this apparent permanence is anything but permanent! Contrary to what we see in science fiction movies and books, a spacecraft is not free to move as it likes once it is beyond the barrier of Earth's atmosphere. It can't just turn on the spot or come to a dead stop in space. Behind its apparent tranquillity, the Earth keeps a firm hand on anything in its vicinity as a result of the force of universal gravitation to which all celestial bodies are subjected.

Any object in space – be it a satellite or probe, or even a planet, comet or asteroid – is a projectile travelling at breathtaking speed on a continuous and inescapable path. This path is called its orbit. Get the circular orbital velocity right of a Geo craft and the satellite's altitude is amost secured. There can be minor gravitational influences, including a very real pull from mountain ranges like the Alps and Himalayas as well as the Moon. Satellites are also affected by solar activity and as a consequence drift slightly away from their ideal position.

If these combined elements were left unchecked then after about 15 days or so the satellite would be operating beyond its pre-destined orbital slot. Viewers would not be able to see their channels.

Using ground-based telemetry, technicians maintain station-keeping adjustments, firing for a moment or two the various thrusters positioned around the satellite's main core to bring it back onto station. These movements can be either in the North-South plane, or East-West, or both. It isn't simply enough for the engineers to do their job – challenging as it is – they have to do it with scrupulous attention to the amount of on-board fuel needed to complete the manoeuvre.

A satellite is lofted into space with a finite amount of on-board propellant, normally between 2 and 3 tonnes, and usually enough for 17 or 18 years of life, and thus guaranteeing on-station performance for about 15 years, but with a little in reserve for emergencies.

There's one other major headache for the satellite's owners and technicians, and this concerns Eclipse Protection. All geostationary satellites can handle the period of time when they are in the shadow of the Earth and thus not able to use the Sun's radiated light and energy for its operation. Geostationary craft use on-board batteries, charging them when in the sunlight, and using that on-board electricity during the craft's 'night'.

1987

Terry Waite captured.
Kept hostage until 1991

TV Towers: How High is High?

Geoff Bains in his 'Location, Location, Location' chapter paints a fascinating picture of a communication satellite operating at a height of "110,000 Eiffel Towers" above the Equator. We spoke to Friedrich von Borries, who is a professor of design theory at the Hamburg University of Fine Arts (HFBK). He was the curator of a very successful exhibition: 'Television Towers – 8559 Metres of Politics and Architecture', at the German Museum for Architecture. We asked him what it is about TV towers and whether people still need these towers now that we are in the satellite age.

"In truth, fewer and fewer television towers are needed. They are admittedly becoming superfluous with satellite television, even if, for example, digital terrestrial television is still transmitted from the Berlin TV tower. In addition, people fly a lot; they are familiar with the view from above – that was entirely different in the sixties and seventies."

"However, a television tower is always more than just a tower. It is a construction of great political symbolic power. Television towers are always higher than technically necessary; they often have rotating restaurants, which are really not required for transmitting television signals. Television towers were also regarded as touristic magnets, as representative edifices in order to show off and thus became more and more expensive and complex. They had to be built for transmission but since they were meant to carry a high load anyway, it seemed natural to invest more money. We cannot help but feel their fascination, for example, in Berlin where the television tower dominates the cityscape of the East. This tower with its glittering ball no longer stands for the dark sides of the German Democratic Republic (GDR) but rather for a positive belief in the future."

Prof. von Borries says the Eiffel Tower was an archetype for many people. "The tower in Tokyo, which is a copy of the tower in Paris and therefore a sign for Japan's opening towards the West, stands for an era as well. Then there is the tower in Stuttgart which was initially meant to become a chimney with a transmitter but instead became a landmark for the city of cars and the enthusiasm for engineering technology. Now new forms are trying to establish themselves – for example in Shanghai and Guangzhou, which deliberately stand for the Chinese tradition and are not meant to be a Western imitation. It always seemed logical to turn such an expensive structure – which functioned as an important device for transmitting the TV signal for a long time – into a political statement as well."

"Today, the struggle for prestige has shifted from TV towers to the construction of skyscrapers. A building such as the Burj Dubai (now the Burj Khalifa) is the translation of the TV tower into another time. 40 years ago they would surely have built a tower. But in times of the internet and Twitter, television is no longer a manifestation of power and communication has diversified. Now you need conference halls, hotels and entertainment. A forerunner here is the TV tower in Las Vegas, which is not a TV tower at all, but the envelope for a hotel and offers a casino."

Asked which TV tower was his personal favourite, he said: "I am going to stick with the Berlin TV tower because I got married there! The Berlin tower has changed from a political structure to an overall symbol for Berlin. Today it is a spaceship of the party culture."

It is difficult to imagine how high 36,000 kilometres really is. It is about 110,000 Eiffel Towers piled up on top of each other.

Kuwait Tower (187 m)

Eiffel Tower (324 m)

Moscow's Ostankino Tower (540 m)

87/06/11
Margaret Thatcher wins third election

87/07/04
Klaus Barbie, Gestapo chief, sentenced to life

Emley Moor, in Yorkshire, is one of Europe`s tallest
TV towers, at 330.4 metres (1084 ft). It replaces
a 386-metre tower that collapsed in 1969.

03

A Place in Space

03
A Place in Space

Geoff Bains

Sir Arthur C. Clarke (1917-2008):
sci-fi writer, inventor, futurist, polymath

It was then a far-sighted idea, but one that has since been made real in the fleet of Astra satellites that today bring television, radio and multimedia services to more than 100 million households across Europe.

87/08/31
Michael Jackson releases 'Bad' album

C. Forrester (ed.), *High Above: The Untold story of Astra, Europe's leading satellite company*,
DOI 10.1007/978-3-642-12009-1_3,

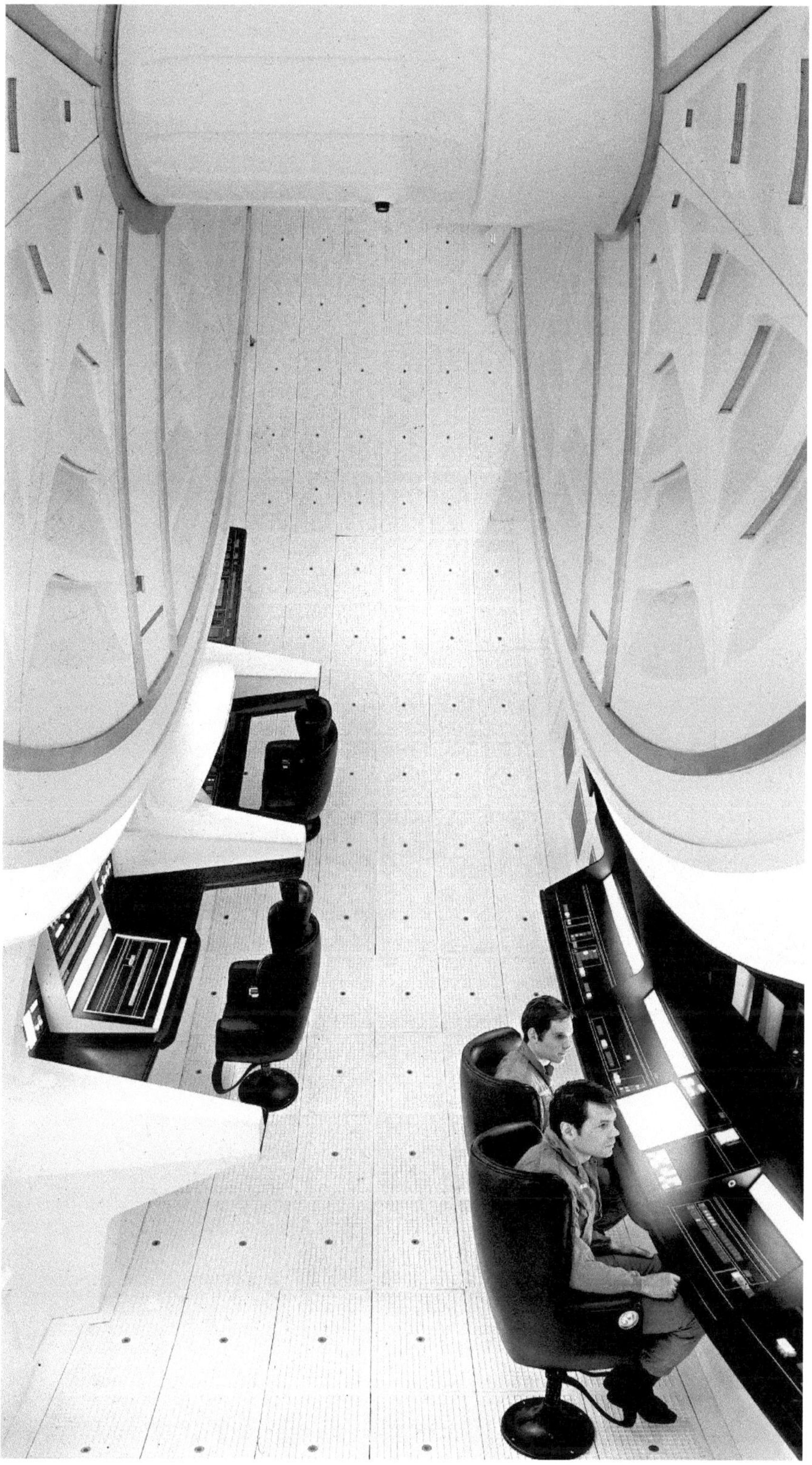

A scene from '2001: A Space Odyssey'

Who were the original thinkers in satellite?

Sir Arthur C. Clarke is perhaps best-known around the world for his science fiction writings, his TV programmes, and most of all as the writer of the seminal 1968 Oscar-winning film, '2001: A Space Odyssey'. But prodigious, varied and credible though his creative output has been, for many, his key achievement was as the founding father of satellite communications.

Although directly involved in neither its scientific nor commercial development, Clarke's greatest effect on the people of the world has been through satellite television, for he is the man who "started it all". In 1945, Clarke wrote of his vision of communications satellites circling the globe in the geostationary orbit, bringing everyone together in truly global communication. It was then a far-sighted idea, but one that has since been made real in the fleet of Astra satellites that today bring television, radio and multimedia services to more than 100 million households across Europe.

All aspects of Clarke's work brought honours on him, including the award of a British Knighthood in 1998. Perhaps the most fitting tribute is that the geostationary orbit, in which telecommunications satellites now work their wonders according to his vision of over 60 years ago, is known – both officially by the International Astronomical Union, and colloquially by almost anyone interested in the subject – as the Clarke Orbit or Clarke Belt.

Arthur Charles Clarke was born in Minehead, on the English Southwest coast, in 1917. His father had been a Post Office engineer but turned to farming, and died when Clarke was just 13. In between the demands of school and working on the family farm, Clarke found time for astronomy, rocket-building, electronics, and reading science fiction magazines, and at 16 he joined the British Interplanetary Society (BIS).

Arthur C. Clarke at home in Sri Lanka

Arthur C. Clarke's successes – and failures!

Sir Arthur C. Clarke has been called many things: visionary, inventor, author, TV star, technician and undoubted futurist. While not the originator of the concept of geostationary satellites (that honour goes to Herman Potočnik in 1928), it was Clarke's additional work, published in Wireless World, which was a case of the right text at the right time. An undoubted success.

His work as a science fiction writer gave Clarke a worldwide audience. Perhaps his most famous work was '2001: A Space Odyssey' (a novel and film script based on an earlier work of his, 'The Sentinel', published in 1948). '2001' itself was published in 1969. The Stanley Kubrick-directed movie came out in 1968. Arthur C. Clarke mostly got '2001' wrong in that computers, useful as they are, do not – quite – have human-like intelligence.

Any sci-fi writer can occasionally get a few things wrong. One of his best-selling books, 'A Fall of Moondust' (1961) has a group of space tourists on "Selene" incarcerated in a sea of choking lunar dust. "Selene" travels through the dust like a submarine. It was a great yarn, but technically nonsense. But given that the Apollo 11 first human landings on the Moon didn't happen until 1969, and there were anxieties about the amount of dust on the Moon, it was a good idea.

Clarke left school in 1936 for a job in the Civil Service, in the audit department, in London. Here he started to write for the BIS journal and for other magazines, exploring the future of space technology as he saw it.

The outbreak of the Second World War interrupted Clarke's spare-time career although his day job ensured that his call-up was deferred. However, in 1941 Clarke joined the Royal Air Force and his interests and abilities in technology earned him training in electronics for the development of "Radio Direction Finding". This was the top-secret development of radar, and before long, Clarke had full responsibility for maintaining the prototype apparatus.

As the War came to a close, Clarke, and other members of the BIS, spent much time wondering how rockets could be exploited commercially so that their development could lead to space exploration. Towards the end of 1944, Clarke came up with the idea of using satellites in geostationary orbit to provide worldwide communications links.

Clarke did not invent or 'discover' the geostationary orbit. That a satellite of Earth at the correct altitude of about 36,000 kilometres completes one orbit in 24 hours – the same time that the Earth beneath it takes to complete one revolution – and so appears to be fixed in the sky was well known to scientists and mathematicians back to Isaac Newton. However, Clarke had the vision to see that in this unique orbit, satellites could be used as relay stations for coverage of vast areas of the world's surface.

He first proposed the idea in a letter in February 1945 to 'Wireless World', a UK radio enthusiasts' magazine, advocating the use of German V-2 rockets in an "immediate post-war research project" into the ionosphere, ending with "an 'artificial satellite' at the correct distance from the Earth wich would ... remain stationary above the same spot and would be within optical range of nearly half the Earth's surface. Three repeater stations, 120 degrees apart in the correct orbit, could give television and microwave coverage to the entire planet."

Herman Potočnik

1988

88/01/01
Perestroika initiated in Russia by Gorbachev

In his letter, Clarke predicted that these developments could be achieved "perhaps half a century ahead". As it turned out, the first geostationary communications satellites were in place within just 20 years. Indeed, 50 years later, Astra was orbiting its fifth television satellite, Astra 1E, for new all-digital TV services across Europe.

He wrote up the idea at greater length in May 1945 for circulation among friends and fellow BIS members, and in July 1945 produced his now-famous article for the October issue of 'Wireless World'. There, Clarke proposed the use of three space stations in geostationary orbit, each to receive radio signals from Earth and re-transmit them back to the ground or on to the other stations, providing coverage over the entire planet.

"I suspect that my early disclosure may have advanced the cause of space communications by approximately 15 minutes."

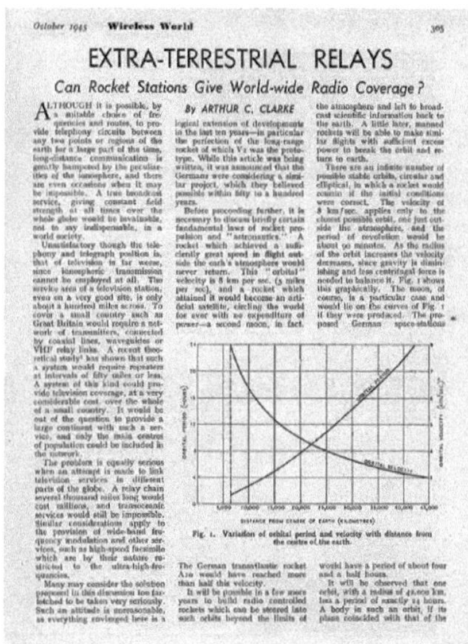

October 1945 **Wireless World** 305

EXTRA-TERRESTRIAL RELAYS

Can Rocket Stations Give World-wide Radio Coverage?

By ARTHUR C. CLARKE

ALTHOUGH it is possible, by a suitable choice of frequencies and routes, to provide telephony circuits between any two points or regions of the earth for a large part of the time, long-distance communication is greatly hampered by the peculiarities of the ionosphere, and there are even occasions when it may be impossible. A true broadcast service, giving constant field strength at all times over the whole globe would be invaluable, not to say indispensable, in a world society.

Unsatisfactory though the telephony and telegraph position is, that of television is far worse, since ionospheric transmission cannot be employed at all. The service area of a television station, even on a very good site, is only about a hundred miles across. To cover a small country such as Great Britain would require a network of transmitters, connected by coaxial lines, waveguides or VHF relay links. A recent theoretical study[1] has shown that such a system would require repeaters at intervals of fifty miles or less. A system of this kind could provide television coverage, at a very considerable cost, over the whole of a small country. It would be out of the question to provide a large continent with such a service, and only the main centres of population could be included in the network.

The problem is equally serious when an attempt is made to link television services in different parts of the globe. A relay chain several thousand miles long would cost millions, and transoceanic services would still be impossible. Similar considerations apply to the provision of wide-band frequency modulation and other services, such as high-speed facsimile which are by their nature restricted to the ultra-high-frequencies.

Many may consider the solution proposed in this discussion too far-fetched to be taken very seriously. Such an attitude is unreasonable, as everything envisaged here is a logical extension of developments in the last ten years—in particular the perfection of the long-range rocket of which V2 was the prototype. While this article was being written, it was announced that the Germans were considering a similar project, which they believed possible within fifty to a hundred years.

Before proceeding further, it is necessary to discuss briefly certain fundamental laws of rocket propulsion and "astronautics." A rocket which achieved a sufficiently great speed in flight outside the earth's atmosphere would never return. This "orbital" velocity is 8 km per sec. (5 miles per sec), and a rocket which attained it would become an artificial satellite, circling the world for ever with no expenditure of power—a second moon, in fact. The German transatlantic rocket A10 would have reached more than half this velocity.

It will be possible in a few more years to build radio controlled rockets which can be steered into such orbits beyond the limits of the atmosphere and left to broadcast scientific information back to the earth. A little later, manned rockets will be able to make similar flights with sufficient excess power to break the orbit and return to earth.

There are an infinite number of possible stable orbits, circular and elliptical, in which a rocket would remain if the initial conditions were correct. The velocity of 8 km/sec. applies only to the closest possible orbit, one just outside the atmosphere, and the period of revolution would be about 90 minutes. As the radius of the orbit increases the velocity decreases, since gravity is diminishing and less centrifugal force is needed to balance it. Fig. 1 shows this graphically. The moon, of course, is a particular case and would lie on the curves of Fig. 1 if they were produced. The proposed German space-stations would have a period of about four and a half hours.

It will be observed that one orbit, with a radius of 42,000 km, has a period of exactly 24 hours. A body in such an orbit, if its plane coincided with that of the

Fig. 1. Variation of orbital period and velocity with distance from the centre of the earth.

Arthur C. Clarke's article on 'Extraterrestrical relays' in July 1945

The space stations were to be manned (operators were required to maintain the valve electronics foreseen in those pre-transistor days) and powered – as are the TV satellites of today – by the Sun, albeit the Sun's heat used to heat small-scale steam engines to generate electricity. Clarke even foresaw the use of on-board batteries for eclipse protection so the satellites could remain functional 24 hours a day.

No enthusiastic response greeted the publication of Clarke's 'Wireless World' article and he later related how the whole idea seemed to be met with "monumental indifference". However, while the article went unnoticed in most quarters, Clarke's proposition struck a chord with the U.S. Navy, which at that time was investigating the use of rockets to improve communications to the US fleet across the World. As a contribution to the U.S. Navy's programme, Clarke's article helped to kick start the development of communications satellites in the 1950s and 1960s.

Clarke did much himself to drive the development of global communications. Often called upon to deliver speeches at political or scientific gatherings, he always advocated his ideas for unifying mankind by globalising communications.

Clarke always denied his position as the father of satellite communications. He has grudgingly acknowledged a peripheral, godfather role but is quite unassuming about the whole affair. In 1982, in his acceptance speech for the Eighth Marconi International Fellowship Award, Clarke commented, "If I had not proposed the idea of geostationary relays... half a dozen other people would have quickly done so. I suspect that my early disclosure may have advanced the cause of space communications by approximately 15 minutes."

It's a typically self-deprecating belief from a man who has some mastery of both science and the arts, who through his own work and the work of others following his vision has brought pleasure to countless millions.

In 1956 Clarke moved from England to Sri Lanka, to follow another passion, scuba diving and underwater exploration. Struck by post-polio syndrome and confined to a wheelchair in 1995, he continued to write about the unfolding future and the benefits of space communications to all mankind until his death, aged 90, in 2008.

In a 1982 'Science Digest' interview, Clarke said, "If you are an optimist, you have a better chance of making a self-fulfilling prophesy. If you say this is a wonderful world and we can make it better, then there is a chance that people will listen to you and do what you say."

The world certainly listened to Sir Arthur C. Clarke and today communications satellites are performing the everyday miracles he first described over 60 years ago.

The world certainly listened to Sir Arthur C. Clarke and today communications satellites are performing the everyday miracles he first described over 60 years ago.

88/06/08
Rupert Murdoch announces four Sky channels on Astra

88/06/11
Nelson Mandela 70th birthday concert, at Wembley

John Pierce – father of satellite technology

An alternative contender for the title of 'father' of satellite TV, and certainly the major individual contributor to the technology used, is John Robinson Pierce – another scientist who also wrote science fiction stories, although Pierce devoted his working life to science and engineering and only wrote science fiction (under the pseudonym of JJ Coupling) as a hobby.

Unaware of Clarke's 1945 article for 'Wireless World', Pierce wrote a paper on the possibilities of communications satellites, which was published in 1955. Pierce discussed several ideas for unmanned communications satellites, including passive reflective balloons and active craft containing receivers and transmitters. Pierce suggested using satellites at low altitudes and in geosynchronous orbits, and even the use of Earth's gravity to control the orientation of a satellite.

But Pierce's contribution to the development of satellite TV was not just limited to academic papers. Born in 1910 in Iowa, USA, Pierce studied electronic engineering at Caltech, gaining his PhD there before taking up a job at AT&T's Bell Telephone Laboratories in 1936, eventually to become Director of Research.

Pierce's first work for Bell Labs was on vacuum tubes and he established the mathematics behind the travelling wave tube, invented by Rudolf Kompfner in a British radar lab, later saying, "Rudy Kompfner invented the travelling wave tube, but I discovered it." Travelling wave tube amplifiers found use in high gain wide-band amplifiers and form the basis of the transmitters in all communications satellites today.

Another early success was leading the team that invented the transistor in 1947, and Pierce himself first coined the name for the new device that was to revolutionise electronics.

In 1958, with the Bell Labs team investigating the potential of communications satellites, Pierce learnt that NASA was proposing developing large balloon satellites for investigating air resistance. Pierce persuaded NASA to change the course of the project and in 1960, Echo 1 was launched – a 30-metre metallised balloon in low Earth orbit to reflect radio transmissions from the ground. Echo 1 was used in the first-ever US coast-to-coast TV transmissions and remained in orbit performing tests for eight years.

However, Pierce was already more interested in active communications satellites. With the success of Echo 1 as impetus and the many departments of Bell Labs providing the expertise (including both the transistor and TWTA), Pierce's team produced the first active communications satellite, Telstar, launched in 1962.

Pierce left Bell Labs in 1970 to return to Caltech as Professor of Electrical Engineering, later also taking the position of Chief Engineer at the Jet Propulsion Laboratory. In 1980 he moved to Stanford's Centre for Computer Research in Music and Acoustics. Pierce retired from Stanford in 1998 and died in 2002, aged 92.

Steve Birkill – first amateur reception

In 1975, Stephen Birkill became the first person to receive satellite TV programmes from a home dish. With a dish system he constructed himself, Birkill astounded the broadcasting world by watching the SITE educational transmissions to rural India over NASA's ATS-6 satellite, at his home in Sheffield, England.

Birkill was born in Barnsley, England in 1946 and became interested in long-distance TV reception at the age of 16, erecting an 18-metre rotating aerial-laden mast through the roof of his parent's home.
In 1966 he joined the BBC, working with, and eventually managing, the transmitter maintenance teams.

In 1975, after hearing of the educational transmissions in India, Birkill constructed his dish from chicken wire, built an LNB from scratch, and put together a receiver from both homemade and commercial modules. News of this accomplishment spread (not least because Birkill was featured on several news and technology programmes produced by his employer, the BBC) and Bob Cooper in the USA invited him to a 1978 trade conference to introduce him to like-minded US experimenters.

In 1981, Birkill left the BBC to start his own business developing equipment for the imminent satellite TV revolution in the UK. He produced communal TV receivers and (with UK electronics company, Ferguson) receivers for satellite-delivered cable TV in the pre-DTH UK before Astra. Birkill also founded Real-World Technology to design high performance reception equipment. The company provided signal monitoring services for UK cable broadcasts in the 1980s and designed the satellite tuners used in all early Amstrad receivers, including the Amstrad SRX-100, the first UK satellite receiver designed for Sky TV's broadcasts on the new Astra satellite.

88/08/18
George H.W. Bush nominated for President

88/09/17
Summer Olympics open in Seoul, South Korea

Bob Cooper – 'Inventor' of home satellite TV

American, Bob Cooper probably did more than any other individual to turn satellite TV from delivery of television signals to local broadcasting centres into distribution of the channels directly to viewers in their homes, initiating the industry where Astra plays an important role. Cooper led and galvanised the 'back yard' experimenters and garage inventers of the USA in the 1970s and 1980s to create the first home satellite TV industry.

In the 1970s, Cooper was in Oklahoma City running a communal TV equipment installation company and working as a journalist specialising in amateur radio and TV technology. Cooper had been experimenting with long distance TV reception since he was 12 years old.

He began publishing the Community Antenna Television Journal (CATJ), full of technical and practical information for the communal TV industry. In 1979, Cooper campaigned against the US Federal Communications Commission (FCC) regulations requiring a licence for C-band reception dishes (only Cooper and Ted Turner held FCC licenses for a private home), and later that year the licence requirements were dropped, opening the doors for legal home dishes.

Cooper was always a firm believer in free-to-air TV and after moving outside of US jurisdiction to the British West Indies in 1980, he campaigned against the introduction of encryption to US satellite TV. In 1990, Cooper moved to New Zealand, and published the monthly 'SatFacts' newsletter, still pushing the boundaries of what could be done with a home satellite dish, both technically and politically. In 2008, Cooper, now 70, finally retired from active satellite TV campaigning, with the knowledge that home satellite TV reception is now commonplace, with dozens of satellites such as those from Astra, beaming TV channels to hundreds of millions of homes across the planet.

88/10/11
One day late, Astra 1A is launched from French Guiana

88/10/30
Ayrton Senna wins Formula One championship

88/12/02
Benazir Bhutto is made Prime Minister of Pakistan

04

Luxembourg: The Mouse that Roared

04
Luxembourg: The Mouse that Roared

It is difficult today to look back 25 years and remember what a major shock satellite television was to the established broadcasting status quo across Europe. Public broadcasting, much of it good, had an entrenched position while emerging commercial networks were already shaking up the industry. But satellite would provide the breakthrough, the electric shock to a system that had developed over three decades in a rather slow, stable and state-controlled manner.

And that shock came from the heart of Europe and yet from one of its tiniest member states: Luxembourg.

The fact that a global satellite TV company should emerge out of one of Europe's smallest countries may sound like a strange alchemy. Several factors turned out to be the perfect ingredients to drive the project forward: Luxembourg's tradition for cutting edge broadcasting technology, coupled with a strong entrepreneurial spirit plus the political will to take risks and look for innovative technologies, as well as the financial backing of several of the country's banks. The determination of the founders and supporters and the enthusiasm of its staff turned a pioneering and financially risky venture into Europe's first commercial and at the same time most successful direct-to-home (DTH) satellite TV company: the Société Européenne des Satellites (SES), operating under the brand name "Astra".

Luxembourg was certainly not a media no-man's land; a liberal audiovisual policy gave it a long and effective broadcasting history, and one that has encouraged the nation to consistently punch well beyond its weight.

This history dates back into the early thirties when, unlike its European neighbours, the small country decided against creating a publicly-funded radio system, instead granting the right to exploit the radio spectrum

1989

89/01/07
Astra 1A arrives at its orbital slot, 19.2 degrees East

89/01/24
George H. W. Bush succeeds Ronald Reagan

C. Forrester (ed.), *High Above: The Untold story of Astra, Europe's leading satellite company*, DOI 10.1007/978-3-642-12009-1_4, © Springer-Verlag Berlin Heidelberg 2011

allocated to the Grand Duchy to a commercial entity. In 1931, the Compagnie Luxembourgeoise de Radiodiffusion (CLR) was born, with French and Belgian banks as well as Havas and Schlumberger as investors. The young company took its headquarters in the 17th century fort 'Villa Louvigny' in the centre of Luxembourg. In 1933, Radio Luxembourg started broadcasting in Long Wave from a brand new and hugely powerful 2500 kWatt transmitter in the East of Luxembourg, at Junglinster.

In 1936, the famous RTL studio in Rue Bayard, in the 8th arrondissement of Paris, was set up. Suspended during the war and the German invasion, Radio Luxembourg recommenced broadcasting in 1945 and using the Medium Wave at 208 metres, achieved significant recognition as an English-language broadcaster of popular music in the 1950s.

At much the same time, television broadcasting started. The first European TV event – with very few viewers, however, at that time – was the special transmission of the crowning of Queen Elizabeth II in London in 1953. CLR started regular television in 1955. One of the early fortunate TV viewers was Luxembourg's royal family which lived at the Château de Betzdorf at that time and had a Philips TV set which had been donated by the Dutch Queen Juliana in 1956.

With the launch of television, CLR had become CLT (Compagnie Luxembourgeoise de Télédiffusion). This group expanded rapidly with the liberalisation of the television sector in the eighties. Partnered with the media company Bertelsmann, it launched RTL in Germany and went on to help create M6 in France, continuing its growth across Europe into the 1990s. With a significant expansion in production, the group developed into RTL Group, Europe's largest commercial television and radio broadcaster and one of the most important TV producers in the world.

Her Majesty Queen Elizabeth II rides in her golden state coach on June 2, 1953. The BBC set up its biggest-ever outside broadcast.

89/02/01
Astra starts transmitting 'FilmNet'

89/02/02
Sky Television launched on Astra

89/02/14
Iran's Ayatollah Khomeini outlaws Satanic Verses book

89/03/24
Exxon Valdez oil spill

More than Steel

The economic success of Luxembourg in the industrial age had almost exclusively been based on steel, with the state-controlled Arbed (today's ArcelorMittal) as its major player. When that industry lost momentum and became more and more commoditised, it represented a growing risk for the country to depend on it as a major employer. The Grand Duchy pushed hard to find new areas of growth and future success. Since the late 1970s, it had started to develop the banking sector into one of its new economic pillars. At the same time, it decided to put a second focus on the media and high-tech sector and use state licenses in broadcasting and space to nurture RTL's growth and international expansion.

"What was important was the sense that something new had to happen. A large country does not have to take as many risks as a small country", says François Biltgen, today Luxembourg's Minister for Communications and Media. "We were in the right place at the right moment."

In 1977, under the umbrella of the International Telecommunications Union (ITU) the World Administrative Radio Conference (WARC) was convened in Geneva. At its conclusion, a formal plan was adopted giving each country in Europe, Africa and Asia – including Luxembourg – assignments at an orbital location with, in general five Broadcast Satellite Services (BSS) satellite frequencies. The countries in North and South America decided to wait until future developments in technology had evolved before agreeing to back the plan.

A satellite of its own suddenly seemed for Luxembourg a desirable and attainable goal. However, the consensus at the WARC-77 had been that the use of the frequencies for television would in any case be limited to the national borders of each country using them. In other words: 'television within frontiers'. The allocations also required the highly contentious use over Europe of so-called Direct Broadcasting by Satellite (DBS) orbital slots and frequencies, and demanded very tightly

"Astra slipped through a loophole, and got away with it."

Behind the scenes at the World Administrative Radio Conference (WARC)

The International Telecommunications Union (ITU) is an affiliate organisation of the United Nations. Originally formed in 1866, it deals with all international matters relating to radio frequency use. Amongst other things, it defines radio services (such as the Fixed-Satellite Service and the Broadcasting Satellite Service) and in these two cases had adopted "Plans" that either assign frequencies or allot orbital positions to national administrations for their use.

The "Plan" adopted at WARC-77 assigned 984 channels within 253 beams to cover the territory of 131 countries.

In 1983, the countries of North and South America met at RARC-SAT'83 and adopted a "Plan" which assigned 2,587 channels at 70 orbital positions to 38 countries.

89/04/21
Student protests in Tiananmen Square

89/05/01
Disney-MGM Studios open

Luxembourg's original broadcasting claim to fame lay with its giant radio transmission towers at Junglinster. While not terribly high (216 metres), they were immensely powerful, beaming longwave transmissions into Germany and France, and in the early 1950s, mediumwave broadcasts into Britain (208 metres)

89/06/03
Ayatollah Khomeini dies

89/07/31
Nintendo releases the Game Boy portable

focused satellite beams. The British footprint ceased sharply at the English Channel, while the French DBS signal was equally tightly drawn, physics being physics, the French scheme had overlap into French-speaking Switzerland, Belgium's Wallonia and mostly French-speaking Luxembourg. Luxembourg obviously would potentially find itself in a nearly absurd situation: the country's overspill into Germany, France and Belgium would have been significantly larger than the country's tiny land mass itself.

With Germany's TV-Sat, France's TDF satellite and UK's BSB, Luxembourg's big neighbours started to aggressively develop their satellite projects. The Grand Duchy pursued its plans in space through a vehicle called LuxSat, a project managed and driven by RTL. The plan was to develop a high-power satellite with a very limited number of transponders serving for the transmission of television and bring it into orbit by 1985. The project was strongly backed by the Luxembourg government and its Prime Minister Pierre Werner. The PTT Ministers in Paris and Bonn were looking at their little neighbor's big ambitions in space with increasing displeasure.

But besides these political concerns, LuxSat had also other challenges, including home-made drawbacks. There was the obvious problem of restricted areas of coverage. In addition, there was political pressure on CLT coming from its French shareholders, especially Havas and its president, André Rousselet, who argued strongly in favour of the Group aligning itself with the French TDF project and was opposed to the emerging competition from the courageous neighbour. CLT had to navigate some very choppy political waters!

In March 1982, Luxembourg's Prime Minister Pierre Werner set a deadline for CLT to inform the government if the company was intending to bring its LuxSat project forward and use the valuable frequencies. If not, they would be given to an alternative licensee. Werner's request and a similar initiative from the Chamber of Deputies (National Parliament) in early 1983 did not generate a proper reply. In February 1983, and despite the urgency, CLT did not even put the question on its board agenda. Pierre Werner concluded that CLT no longer had an exclusive right. The game was now once again open for another player to come on the scene and develop its own, alternative plan for Luxembourg's own satellite.

Transatlantic Help

It was at this point that Pierre Werner turned to an American expert who proposed using a new satellite design which had already proved itself in the United States: the famed 'medium-power' satellites. On a visit to Washington for a World Bank conference, he met with Clayton Whitehead. Whitehead was a former presidential advisor who worked for the aerospace company Hughes.

Hughes had developed medium-power satellites, as Whitehead described them. Their advantage was that they needed less transmission power per channel so that the payload could carry a far greater number of transponders: 16 transponders compared with 5 on a high-powered satellite. At the same time, thanks to the technical progress made in terms of reception equipment, the signals transmitted by these satellites could still be received using small-sized satellite dishes.

Another major advantage of the fresh thinking was to use frequencies in the so-called Fixed Satellite Services (FSS) band enabling circumvention on the geographical and regulatory restrictions established in the Broadcasting Satellite Services (BSS) band. Using the FSS band meant placing TV signals onto telecommunication frequencies, and all over Europe: Technically the idea presented little challenge, but a fresh and highly controversial view of frequency use.

"Astra was simply not allowed!" says Christian Schwarz-Schilling, the German PTT minister between 1982-1992, bluntly. And Peter de Selding, a highly regarded industry journalist and commentator, sums up the situation: "Astra slipped through a loophole, and got away with it."

Transatlantic gift: The Luxembourg royal family receives a generous gift from the USA, while at the Château

Whitehead came to Luxembourg at the end of 1983. "The American had views on the future of telecommunication systems that I could hardly believe", remembers Pierre Werner in his memoirs. The new system was called GDL, for Grand Duchy of Luxembourg, with a research company attached to it, called Coronet. In November 1983, GDL-Coronet was formally notified by the Luxembourg authorities to the ITU. Luxembourg's satellite ambitions became obvious. The neighbouring countries remained highly sceptical.

Opposition from Neighbours

In 1984, CLT announced that it had reserved two transponders on the French TDF satellite and said that it would do everything "both legally and politically" to ensure that the GDL project failed.

At the same time, new opponents stimulated the broadcasting scene, in particular Eutelsat – today a privately held company, but at the time an intergovernmental organisation (IGO) of the various European post office and telecommunications administrations, which was modelled on a similar operation, Intelsat, another IGO.

Eutelsat's Director General at the time publicly claimed that GDL was going to cause "significant economic harm" to Eutelsat, and called upon all signatories of the Eutelsat Charter to boycott GDL. In the end it was only Luxembourg's threat of referring the matter the European Court of Justice that finally made Eutelsat abandon this course of action.

Eutelsat's opinion was shared by many, not least by French PTT Minister Louis Mexandeau, who declared that his country was not going to allow Luxembourg's "'Coca-Cola satellites' to undermine our linguistic and cultural identity." Referring to the fact that Whitehead was largely looking for US investors for the new GDL venture, Mexandeau argued: "If the American businessmen attempt to test our abilities to accept their challenge then we will answer them." There were even serious suggestions that any such incoming signals could be 'jammed'.
All of this heavily fuelled the debate in the Grand Duchy. In the run-up to Luxembourg's parliamentary elections of 1984, the situation at GDL and its political as well as financial risks quickly became a topic in the electoral campaign. The leader of the political opposition went as far as denouncing the project from the Speaker's Desk of the National Parliament as a "danger for Luxembourg, and a danger for Europe".

These pressures, and the seemingly non-stop negative press which the project generated, made it increasingly difficult to complete the capital structure of the future operating company and to assemble the necessary funds for financing its first satellite. Whitehead left Luxembourg in 1983. By mid 1984, the GDL project had come to a standstill. And with it – again – Luxembourg's plans and ambitions for its own satellite.

Singer Heintje receives the 'Golden Lion' from Radio Luxembourg

"I was fascinated by the idea that from Luxembourg, we could throw up a signal 36,000 kilometres and reach tens of millions of people. It was like the eighth wonder of the world."

This was the moment for a new player to step in – a new Prime Minister and a new thrust for the project. After winning the 1984 elections, Pierre Werner's successor Jacques Santer remained as committed as his predecessor to the idea of a satellite business. Santer was convinced that Luxembourg needed fresh ideas, new technologies, forward looking investment and engagement. "A small country – but a big laboratory", he said. "I was fascinated by the idea that from Luxembourg, we could throw up a signal 36,000 kilometres and reach tens of millions of people. It was like the eighth wonder of the world."

Santer did not hesitate very long. On the very day of his formal appointment by the Grand Duke, on Friday, 20 July 1984, he held a crisis meeting of ministers at his private home to discuss the issue. "We could have paid off the existing debts of 74 million francs (about 19 million Euros) and given up the project. Or we could carry on and create something unprecedented. Questions were raised about the risks for Luxembourg. But there are times when, as Prime Minister, you stand alone and you have to listen to your own conscience. I had personal belief in the satellite project. Luxembourg is a small country, but I have always been convinced that we have great potential. But to realise that potential, we must be one step ahead of the bigger countries."

Santer succeeded in forging an alliance to support and finance the satellite project.

However, the crisis forced the government to recognise that the Luxembourg state would have to play an even more active role in order to bring together a group of investors. Santer quickly found two public Luxembourg financial institutions, Société Nationale de Crédit et d'Investissement (SNCI) and Banque et Caisse d'Epargne de l'Etat (BCEE), as well as two German banks, Dresdner and Deutsche Bank, which supported the project through their Luxembourg affiliates. These banks were ready to become founding shareholders in the new satellite company. A number of other banks and investors were also asked to join the financial alliance.

One of the earliest and most enthusiastic backers of the SES project was a long-time diplomat and EC Ambassador to the USA in the early 1980s, Count Roland de Kergorlay, who put up 10% of the original 100 million dollars investment for SES. His diplomatic skills also played a crucial role in helping the Luxembourg government to 'see off' French attempts to strangle SES at birth, according to his obituary published in the 'London Times' in 2003.

89/10/17
Earthquake hits San Francisco bay area. 67 die

89/11/09
The Berlin Wall is breached. Thousands cross freely

89/11/12
Brazil holds first free presidential election since 1960

The headquarters of SES (above) and Astra (below) in Château de Betzdorf in the countryside of Luxembourg, 25 kilometres from the city of Luxembourg

Société Européenne des Satellites was the name under which the Luxembourg-based satellite operator was incorporated on 1 March 1985. It was later abbreviated to SES. The first SES team decided to use this name as a corporate name and look for a different brand name. Astra was chosen to be the market-facing and customer brand and was therefore applied to the satellites as well as to the consumer proposition for the Direct-to-Home (DTH) TV reception.

When SES grew successfully around its European and DTH core business, it started expanding, initially to the Asian region with an investment in Hong Kong-based AsiaSat in 1999, and later in Scandinavia with an investment in Nordic Satellite AB (later called Sirius), and then, in 2000, with the acquisition of General Electric's Americom and its fleet of satellites. Following this significant expansion and an Initial Public Offering, SES morphed into SES Global, with the Astra business now called SES Astra and the American sister branch called SES Americom, based in Princeton, New Jersey.

Following the acquisition of New Skies Satellites based in The Hague, Netherlands, as a third pillar, the "Global" element was dropped from the corporate SES name. SES New Skies and SES Americom were merged into SES World Skies in 2009. The headquarters of both SES and SES Astra are in Betzdorf, Luxembourg; the headquarters of SES World Skies are in The Hague and Princeton. At the Stock Exchanges of Paris and Luxembourg, the SES company is listed under the SESG symbol.

The Start of SES

On March 1, 1985, the Société Européenne des Satellites (SES) was incorporated and started to work out of the often cramped suite of offices near the main Luxembourg railway station. The company had eleven founding shareholders and barely a dozen employees.

As its name suggests, SES was positioning itself from the start as a European undertaking with exclusively European shareholders. The European character of the company was essential in order to alleviate the anxieties of those who continued to fear that Luxembourg's satellite ambitions were a Trojan horse for US cultural imperialism.

John Tydeman was one of the first members of the new venture. "The press, when I joined the project, was very sceptical. There was no satellite, there were no customers. My situation was that I had spent the best part of a year in New York working for Rupert Murdoch on a similar, but less ambitious project, so I had to say I did not see the Luxembourg project as being high-risk at all. The uncertainty that I saw was the train wreck left after the Coronet element where there were disillusioned shareholders, and a genuine concern as to whether the project would work. Mr. Whitehead and his team had quite brilliantly solved the problem of orbital slots, although there was still much work to be done in clearing frequencies. But in my mind, the concept of DTH television was starting to take off and was by no means a wild concept. Cable systems were being fed by satellite and the technology was changing quite radically and so the notion of ever-smaller dishes and home-based reception was a very real possibility."

On September 24, 1985, SES ordered its first satellite from manufacturer RCA. While Luxembourg's plan proceeded, despite its struggles, its European neighbours had more fundamental problems. Germany's TV-Sat suffered a failed solar panel and there were major commercial concerns about France's TDF project.

Serge Siritzky, editor of the French trade magazine 'Ecran Total', explains the course of events from a Parisian point-of-view: "Perhaps the greatest difference between Luxembourg's SES and France's TDF is that

24 September 1985, an important date for SES: The satellite contract is signed.
Roland de Kergorlay / Vice-Chairman, Corneille Brück / Chairman, Charles A. Schmidt / RCA Astro-Electronics, Div. Vice-President and General Manager

1990

90/01/03
Manuel Noriega surrenders in Panama

Astra's launch VIPs at Kourou, in French Guiana: Pierre Werner and Luxembourg's Grand Duke anxiously await liftoff

"I have always been convinced that we have great potential. But to realise that potential, we must be one step ahead of the bigger countries."

the TDF management had no marketing culture. The French satellites were called TDF1 and TDF2 and the Germans TV Sat1 and TV Sat2. They had the same location in the sky. But, TDF1 was launched in 1989, before having signed any contracts with clients and before really knowing what channel to put onto the satellite. The only logical client was Canal+, which wisely wanted to reach households in areas which were not covered by its terrestrial broadcasts. Canal+ signed a contract with TDF and rapidly achieved 50,000 subscribers through TDF1. Unfortunately the French and German satellites experienced one failure after another and together could only broadcast four signals and it soon became apparent that they had little future."

These difficulties were not without influence on Luxembourg's project: No investor wanted to run risks similar to those in Germany and France. SES was therefore unable to find a banker prepared to lend it the additional capital required for purchasing and launching its first satellite.

Once again, Jacques Santer had to step in. In a dramatic appeal and with his highest personal commitment, he convinced the government and the parliamentary majority to authorise a State guarantee for a sum of 3.6 billion Luxembourg francs, or more than 100 million dollars at the time. From that point on, Santer's political career was inextricably linked to the success or failure of the project.

In January 1986, the pipe-smoking Swiss Pierre Meyrat was hired and arrived to head up the team. Meyrat had built a cable business in Switzerland with remarkable persistence and success, and launched the first pay-TV company Teleclub, where he had met and partnered with Leo Kirch. "When I came to Luxembourg, SES consisted of one office floor with a dozen employees, two technical freaks, an accountant and secretaries", he says. "I first of all had to find a very good commercial and technical director. Only this way could we succeed in building a strong team in our offices, which quickly became too small."

The Importance of Marketing

Meyrat's concept was firmly NOT the "build it and they will come" strategy, but to build the offer by starting from the viewer's point of view. "I looked at the whole question from the end-consumer's perspective from the very beginning. The boom actually took place in the television retail shops", says Meyrat. "In our conversations with the broadcasters, we saw immediately that the only argument that counted for them was that the satellite had reach. It was therefore of the highest importance that we were able to gather programming packages which were attractive for our different target countries. Britain helped here a lot, of course, since Murdoch knew very early that he would only be successful with direct reception. German channels also were very interested due to the slow development of cable."

90/01/22
Communists end 45 years of power in Yugoslavia

90/01/31
The first McDonald's in Moscow, Russia opens

90/02/11
Nelson Mandela released from prison

Marcus Bicknell welcomes Richard Wolf (then of the Children´s Channel)

Meyrat recognised the importance of marketing and laid out a scheme to use electrical retailers and the trade for marketing, using strong industry partnerships to leverage the company's strengths. "A marketing budget to reach 100 million people in Europe would have been immensely expensive", he says. "That is why we had to look for allies."

He found them. The manufacturers and the retailers became the main interface and partners for Astra to spread the good news about the new technology and its advantages. Ease of use was what Meyrat finally aimed for. "I remember going out in the rain with my technicians and told them to show me how to install a dish in the rain on the roof!"

For Meyrat, it seemed that nobody in Europe had much thought about marketing satellites before Astra. "Eutelsat put technical miracles into orbit and said: 'whoever wants can rent them' but did not care at all about the question of how people should receive the service. This was left to those who contracted the capacity. Our idea was that we had to start with consumers. That was what was new with Astra. Nobody had thought about how the signal would get to the viewer, everyone seemed only to care about renting transponders."

Marcus Bicknell was hired to spell these new principles out, as Astra's first Commercial Director. A more unlikely candidate might, at first sight, be hard to find. Here was Marcus, charming, erudite, but with a background that included being manager of rock band Genesis, and then working for CBS and A&M Records in the 1970s. He was passionate about entertainment and new technology – a fascination that he maintains to this day.

"The biggest challenge of Astra in the early days was that it was a step change in communications and television", says Bicknell. "So whereas people had their three or four broadcast channels, all of a sudden we were proposing the choice of 16 channels and then 32 and then 64 and then the hundreds of channels which we now know. With that step change here, there came a tiny company, in a tiny country, not one of the major parts of Europe, and we were combating other major competitors. And at the time in 1986, there was the small matter of a lack of capital: we needed 350 million euros. We had no frequencies, we had no regulatory environment, we had no clients, we had no satellite, we had no programmers, we had no programmes and we had no reception equipment. Otherwise it was totally perfect!"

Bicknell and his tiny team went to work with enthusiasm. "Our primary target was Sky Television and Rupert Murdoch himself was calling the shots. The decision was not something he delegated. And we had competition, because he was being solicited by BSB, the five-channel British satellite and by others. We also forget the Atlantic project which was similar to our own and based in Ireland. And we had to be pretty clever. We

"A marketing budget to reach 100 million people in Europe would have been immensely expensive. That is why we had to look for allies."

"We needed 350 million euros. We had no frequencies, we had no regulatory environment, we had no clients, we had no satellite, we had no programmers, we had no programmes and we had no reception equipment. Otherwise it was totally perfect!"

met Murdoch several times, as well as his team from America and from Britain and we suddenly got to the stage where we realised that we could have a multi-channel deal." On June 8, 1988, Murdoch announced at a high profile press conference the signing of the first four channels on Astra.

"One cheque for 80 million pounds, then worth 180 million euros, for the use of four transponders for ten years and from then on Astra was made", Marcus Bicknell says. "That was the turning point."

But SES still needed to get its satellite into orbit. Astra 1A had been shipped to Kourou, Arianespace's jungle-surrounded spaceport in French Guiana, and was waiting for its launch in December 1988. Nine years after the first successful Ariane launch in December 1979, and 3.5 years after the inception of SES, the first Astra countdown was prepared, with the rocket on the launch pad and an expensive satellite within the Ariane-4 fairing, ready for a scheduled launch on December 10, 1988.

The co-passenger on board the rocket – as Ariane rockets often transport a pair of payloads – was a secret British spy satellite, SkyNet 4B. The Ariane-4 rocket had an extra pair of powerful liquid and solid rocket strap-on boosters that could take the combined loads safely into space.

Anxious Moments

The first launch attempt had to be scrubbed, as there were technical problems with the rocket. "There was a huge tension for us and all our colleagues", remembers Pierre Meyrat. "Three years of work and three years of personal efforts and commitment, and now those who had never seen it before, all of our eyes were fixed onto that rocket which stood there with its huge 58 metres in height." "We went there the first night, and the countdown started, but the engineers quickly recognised that a component on the rocket had to be changed. All its highly explosive fuel had to be drained out of the rocket, the piece fixed and the rocket filled up again."

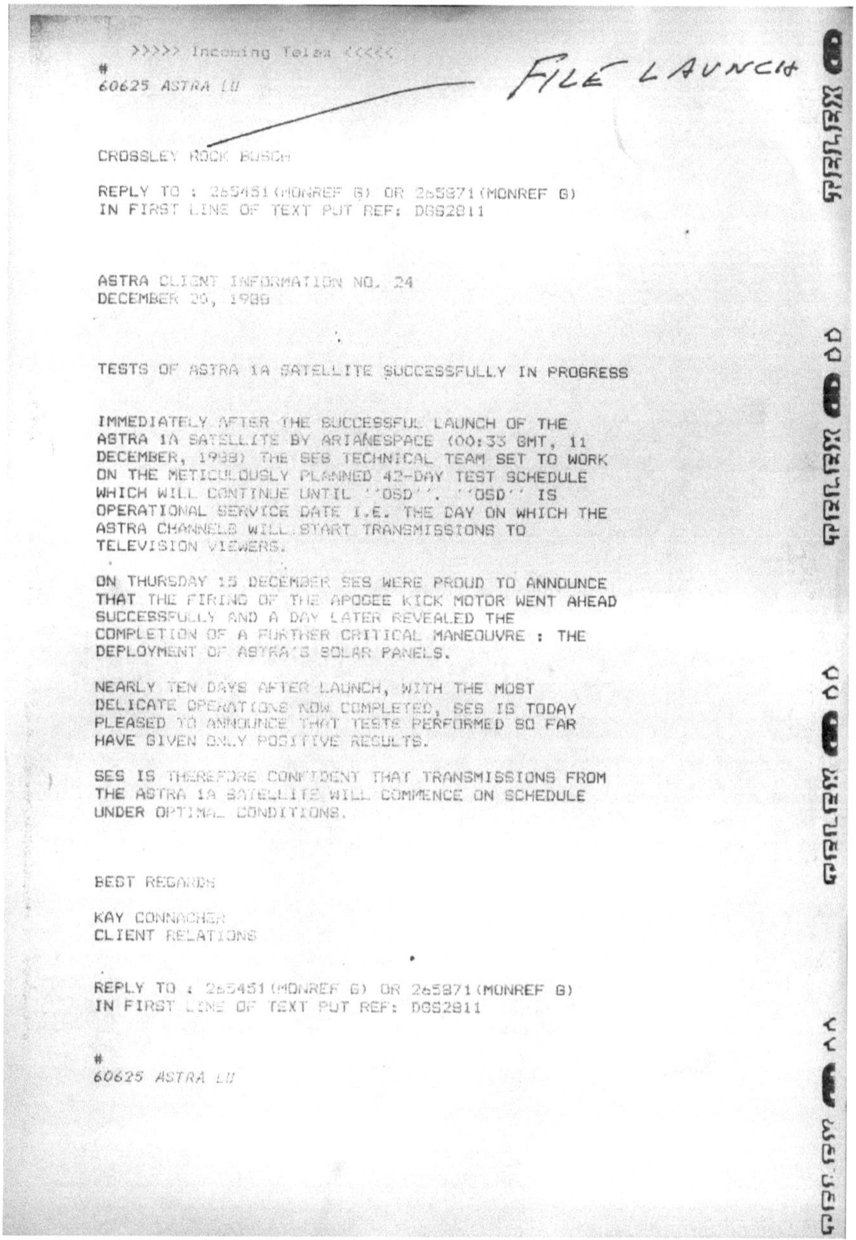

>>>>> Incoming Telex <<<<<
#
60625 ASTRA LU

FILE LAUNCH

CROSSLEY ROCK BUSCH

REPLY TO : 265451(MONREF G) OR 265871(MONREF G)
IN FIRST LINE OF TEXT PUT REF: DGS2811

ASTRA CLIENT INFORMATION NO. 24
DECEMBER 20, 1988

TESTS OF ASTRA 1A SATELLITE SUCCESSFULLY IN PROGRESS

IMMEDIATELY AFTER THE SUCCESSFUL LAUNCH OF THE ASTRA 1A SATELLITE BY ARIANESPACE (00:33 GMT, 11 DECEMBER, 1988) THE SES TECHNICAL TEAM SET TO WORK ON THE METICULOUSLY PLANNED 42-DAY TEST SCHEDULE WHICH WILL CONTINUE UNTIL ''OSD''. ''OSD'' IS OPERATIONAL SERVICE DATE I.E. THE DAY ON WHICH THE ASTRA CHANNELS WILL START TRANSMISSIONS TO TELEVISION VIEWERS.

ON THURSDAY 15 DECEMBER SES WERE PROUD TO ANNOUNCE THAT THE FIRING OF THE APOGEE KICK MOTOR WENT AHEAD SUCCESSFULLY AND A DAY LATER REVEALED THE COMPLETION OF A FURTHER CRITICAL MANEOUVRE : THE DEPLOYMENT OF ASTRA'S SOLAR PANELS.

NEARLY TEN DAYS AFTER LAUNCH, WITH THE MOST DELICATE OPERATIONS NOW COMPLETED, SES IS TODAY PLEASED TO ANNOUNCE THAT TESTS PERFORMED SO FAR HAVE GIVEN ONLY POSITIVE RESULTS.

SES IS THEREFORE CONFIDENT THAT TRANSMISSIONS FROM THE ASTRA 1A SATELLITE WILL COMMENCE ON SCHEDULE UNDER OPTIMAL CONDITIONS.

BEST REGARDS

KAY CONNACHER
CLIENT RELATIONS

REPLY TO : 265451(MONREF G) OR 265871(MONREF G)
IN FIRST LINE OF TEXT PUT REF: DGS2811

#
60625 ASTRA LU

December 20, 1988, and all is well with 1A

The actual launch of Astra 1A took place a day later, on 11 December.

"The second evening, it became exciting all over again, the countdown was interrupted once more, but then the rocket blasted off just half an hour later. There was of course a big cheer. Grand Duke Henri and board chairman Pierre Werner had specifically come to Kourou to see the launch, and in Luxembourg sat Prime Minister Santer and the whole SES crew to follow the launch live."

Santer remembers the evening well. "I remember being very nervous as the launch was prepared in Kourou and I was watching here in Luxembourg at RTL. The countdown ran until a few seconds before ignition and then a technical defect was discovered and the launch was postponed. I can tell you, I did not sleep that night! The following day the defect was repaired and thank goodness the launch proceeded without a hitch."

"Luxembourg! I think it could not happen elsewhere. Luxembourg became the mouse that roared."

"It really was the breakthrough of a new era. We have managed to establish the audio-visual sector as a pillar of the Luxembourg economy, but also to create a high-tech product that we can distribute all over Europe thanks to the directive on television without borders. Without Europe, we could never have launched such a project. Astra is a good example of a high-tech project that has succeeded in the laboratory of Luxembourg."

"Luxembourg! I think it could not happen elsewhere", sums up Marcus Bicknell. "Luxembourg became the mouse that roared."

Charlie Ergen President & CEO, EchoStar

"We saw SES' activity in Europe as an entrepreneurial company that was entering the market to unseat some incumbents in what was still a new industry. In the US at that time the satellite business was also a sub-set of small businesses. We at EchoStar, also looking to enter the business, saw SES as one of the pioneers in how to compete. Another company, called PanAmSat, was also in a similar position. We looked closely at the European experience; looking at if SES could be successful in Europe could we do the same in the US. To some extent we looked upon SES as a role model."

"Later we got to know SES much better, certainly once they had acquired GE Americom in the US. This put SES on the road almost overnight as far as the US was concerned."

"Companies who had business with SES in Europe recognized SES as a dependable, reliable supplier of satellite capacity. They had a great reputation, and people in the industry knew about SES, their history and the growth they'd achieved. It meant they had instant credibility when later they entered the North and Latin American markets. They still to this day have probably the best track record in building satellites, launching them successfully and then operating them successfully. It is a record that nobody else comes close to."

"SES, from our perspective, has a couple of key elements for their success. First and foremost they have always been honest with their customers, treating customers with respect. Secondly, without question they have a track record for technical excellence, and to innovate where necessary. Thirdly, their innovation is well known, for example, their use of spot beams which was practically unheard of a few years ago and making satellite [use] even more economical."

90/07/02
A stampede at Mecca kills 1,426

90/08/02
Iraq invades Kuwait. Start of first Gulf War

Astra 1A sits atop its Ariane launch rocket

90/09/10
19-year-old Pete Sampras beats Andre Agassi to win US Open

90/10/14
Composer Leonard Bernstein dies

05

Achieving the Breakthrough

05
Achieving the Breakthrough

Jacques Santer
Prime Minister of Luxembourg, 1984-1995
President, European Commission, 1995-1999

Jacques Santer
Luxembourg's Prime Minster,
and firm supporter of SES

As we celebrate 25 years of Astra, I am at an age where I can look back objectively. On reflection it is quite clear that my predecessor, Pierre Werner, launched what was a visionary initiative to create a satellite industry here in Luxembourg.

The story began in 1977 when Luxembourg received five broadcast channels as a result of the World Administrative Radio Conference in Geneva. We wanted to exploit this opportunity, so we offered the channels to RTL, whose management team was keen to help set up the LuxSat satellite project. However, RTL's shareholders were not unanimously enthusiastic about the project.

90/11/12
Tim Berners-Lee publishes project called WorldWideWeb

C. Forrester (ed.), *High Above: The Untold story of Astra, Europe's leading satellite company*,
DOI 10.1007/978-3-642-12009-1_5, © Springer-Verlag Berlin Heidelberg 2011

Pierre Werner (1913-2002)
Luxembourg's Prime Minister, and visionary

"Questions were raised about the risks for Luxembourg. But there are times when, as Prime Minister, you stand alone and you have to listen to your own conscience."

So the LuxSat programme stalled and by the time I entered government as Finance Minister, in 1979, Pierre Werner had lost patience. He gave RTL an ultimatum that if it did not make progress by a certain date, we, as the state, would launch the project. Needless to say, that threat was not well received and led to a rupture in relations between the RTL shareholders and Mr. Werner's government.

But Mr. Werner was determined, and he started seeking an alternative solution. Then, on a visit to Washington for a World Bank conference, we met with Clayton Whitehead. Mr. Whitehead was a former presidential consultant who worked for Hughes at the time. Hughes had developed a 'medium-powered' system, as he called it. This involved building telecommunications satellites that performed exactly the same functions as the larger satellites, but which could carry many more channels and would be much cheaper to operate.

So the government gave Mr. Whitehead six months to set up a project. When we announced that an American would be steering the project, there was a swell of hostile opposition to the idea. I remember that some disparagingly named it the "Coca-Cola satellite."

At the time there was no European transnational broadcasting directive, so each country had to focus on its own territory. Well, it is obvious that we could not have a satellite just for Luxembourg. But Mr. Whitehead said something that, as a lawyer, I will always remember. "If the technology works, the legislation will follow." And he proved to be right.

That is not to say that we didn't face problems with the creation of SES. Mr. Whitehead had the technology that worked, but he wanted to follow the American model of venture capitalism in establishing the company. That system was not common in Europe at the time, and so he found it difficult to find investors. And although he was given an extension, Mr Whitehead realised that he could not achieve what he wanted. He left Luxemburg and handed us the keys to the company.

We still faced a big decision, however. On the very day that I was sworn in as Prime Minister by the Grand Duke, on Friday, 20 July 1984, we held a crisis meeting of ministers at my private home to discuss the possibilities. We could have paid off the existing debts of 74 million francs and given up the project. Or we could carry on and create something unprecedented.

Questions were raised about the risks for Luxembourg. But there are times when, as Prime Minister, you stand alone and you have to listen to your own conscience. I had personal belief in the satellite project. Luxembourg is a small country, but I have always been convinced that we have great potential. But to realise that potential, we must be one step ahead of the bigger countries.

VIPs regularly visited SES's Betzdorf headquarters. Here Germany's Chancellor Helmut Kohl drops by and is greeted by (L-R) Jacques Santer, Pierre Werner and Pierre Meyrat

Setting up SES was a big gamble at the time. Without the guarantee agreements I negotiated with the two banks SNCI and the BCEE, we would not have attracted investors. The first investors were from the big German banks that were established in Luxembourg, and then the Luxembourg banks followed.

But that did not mean we had launched a satellite, which would cost some five billion francs. I had to create a law to provide a state guarantee of 3.6 billion francs, which we managed to get passed through parliament. Nevertheless, I was putting my political career on the line. Because despite knocking on the doors of insurers all over the world, the one thing we couldn't get insured was the ignition of the launch rocket, that very first second. If that failed, I would have lost the state 3.6 billion francs.

I remember being very nervous as the launch was prepared in Kourou and I was watching here at RTL. The countdown ran until a few seconds before ignition and then a technical defect was discovered and the launch was postponed. I can tell you, I did not sleep that night. The following day the defect was repaired and thank goodness the launch proceeded without a hitch.

It really was the breakthrough of a new era. We have managed to establish the audiovisual sector as a pillar of the Luxembourg economy, but also to create a high-tech product that we can distribute all over Europe thanks to the directive on television without borders. Without Europe, we could never have launched such a project. SES is a good example of a high-tech project that has succeeded in the laboratory of Luxembourg.

1991

91/01/23
'Seinfeld' debuts on NBC-TV

91/02/21
Margot Fonteyn, ballerina, dies

The Plan Takes Shape

In the early 1980s I was working at News Corporation as the Head of their International New Media Group, a small group exploring media project opportunities for investment. In New York, in 1984, we had been working on a project called Town and Country Television in America, which was an attempt to do a small DTH medium-power satellite project. At some stage, Clayton Whitehead had met up with Rupert Murdoch and this led to me looking at the then Coronet satellite project in Luxembourg.

Common wisdom at the time leaned towards high-power satellites for home reception of television so this was a conceptual breakthrough.

News Corp did not proceed with a potential investment in Coronet because among other things, Coronet was structurally complicated. And they wanted News to pay a turnover tax on the revenue generated. This did not go over very well at all. A further complication was that there did not seem to be much support from broadcasters for the Coronet project at that time.

In mid-1985 I was asked to come and work as a consultant with the Luxembourg shareholding group that was taking over from by then the failed Coronet project, to try to resurrect something from the ashes of the project.

These early days were spent in a fairly chaotic environment in an office in downtown Luxembourg although my task, while challenging, was straightforward: to build a commercial business concept supported by a business plan for a medium-power, direct-to-home satellite television operation. The shareholders were understandably nervous and had many questions as this was not only Europe's first private satellite venture but also a business about which they (the shareholders who were mainly bankers) knew very little.

I remember a board meeting, prior to us having signed the first satellite order, at which I was asked, I think by Count Kergorlay, to go and get some Letters of Intent from broadcasters to show that there was potential interest in this project. Our lawyers and shareholders all agreed that while Letters of Intent or Memorandums of Understanding could be written, they would not be binding legally. I traversed Europe, visiting many media operators. We already had [Scandinavian broadcaster] Kinnevik on board, as they were shareholders. We ended up with eight or nine Letters of Intent from Spain, UK and even France, so this together with the technical and commercial reports influenced the Board to at least sign a first payment commitment for the first satellite, an American partially built GE satellite which became Astra 1A. We were also in early discussions with Ariane, which had commenced launching again after its early setbacks. Even if the first satellite was to be American, the launch vehicle would turn out to be European.

When Pierre Meyrat joined as Director General in 1986, and we had gotten to know each other, I was asked to stay on in a more formal capacity and worked with SES until the end of 1994 as Strategic Advisor on all the key commercial and related aspects of the business, of which there were many. Meyrat started expanding the team to include talented people like the technical team and marketing director Marcus Bicknell.

John Tydeman
Strategic Advisor to SES, 1985-1994

91/03/02
Astra 2B launches from Kourou, 16 extra channels

91/03/15
Four LA police charged with beating Rodney King

91/04/03
Cease fire agreed in Persian Gulf War

91/04/14
Thieves steal 20 paintings worth $500 million from Van Gogh museum

It's also fair to say that while we had just one orbital position and even though not all of the frequencies had been cleared, this actually turned out to be an advantage. The other big satellite operators, Intelsat and Eutelsat, were at that time 'old boys' clubs' whose shareholders were the then national telecommunications authorities (PTTs) of the member countries. They had plenty of orbital positions and quasi-monopoly status, but as you cannot run a DTH business expecting every home to fix half a dozen dishes in its backyard, they lacked commercial focus for satellite broadcasting. Perhaps they were relying on another monopolistic regulatory regime emerging at that time, the European national direct broadcast satellite systems (DBS), to manage satellite reception to the individual household.

Our project started to take shape and it certainly generated a lot of 'noise'. SES wanted to be 'European' but the Italians and French particularly as well as Spanish and Germans, initially, dismissed Astra a path for DTH reception. The satellite regulatory regime at the time compounded the problems for Astra as satellite reception required both an uplink and a downlink signatory; two national telecommunications authorities. SES had one signatory with the Luxembourg PTT but securing a second turned out to be extremely complicated. The role of British Telecom (BT), which after a long period of negotiations eventually agreed to participate and thereby give SES legitimacy, was crucial.

Negotiations with other key broadcasters across Europe were difficult and time consuming but picked up momentum as they realised SES had a satellite strategy which offered coverage, power, back-up (going forward from Astra 1A) and an opportunity to control their own destinies locally. This was something that had not been available in Europe before.

Stimulating the manufacture and distribution of satellite receivers and antennas were key elements in the Astra platform. Some of Europe's major reception manufacturers came along almost as wolves in sheep's clothing lobbying at the European Commission to push Europe on an isolationist digital path forward using a non-standard D (digital) Mac reception system. This required intense lobbying by Astra and its pro-active customers such as News Corporation for had they been successful, Europe would have been denied its undoubted successful digital TV platform

These were just some of the many obstacles Astra faced along the way.

Broadcasting at that time was very much a domain of mainly government entities (with some commercial franchises such as in the UK). So here was one of Europe's smallest countries, the Grand Duchy of Luxembourg with three hundred thousand people or so, with a very eclectic group of shareholders and a start-up company ready to pioneer satellite television as we now know it.

Not one of the initial shareholders knew anything about satellite television. What the shareholders wanted, of course, was the assurance that this business could work – and that's something that, even though we were confident, none of us could guarantee with certainty although we had battled extensively with success on every front – commercial, technical, political and regulatory.

They weren't so much worried about the rocket launcher blowing up or the satellite exploding because those elements could be insured against. What concerned them was whether we would gain any customers to justify the investment.

I must say when at the Astra 1A launch the countdown was stopped on the first night with a few seconds to go, it was a very sombre and sobering affair, although leading up to the successful launch on the second night!

While the contract with BT provided legitimacy and thus security to proceed, the path to a more liberalised television regime (through initiatives such as Television Without Frontiers) and the opportunity to start a satellite television operation inexpensively

91/05/15
Europe ends South African trade sanctions

91/05/16
Édith Cresson elected France's first female Premier

The Iron Lady, Margaret Thatcher, and husband Dennis, visit SES Astra

"My task, while challenging, was straightforward: to build a commercial business plan for a medium-power, direct-to-home satellite television operation."

with the potential of a satellite system to deliver the signals to households was bringing many new entrants into the market.

SES, compared to Coronet, was a very different proposition for News Corp. In the SES world they were a potential customer (through their UK satellite company Sky Television). Sky by then had its first channel up and running and News Corp had thought about its path towards DTH transmission in the UK. They grew from a single channel to a four-channel operation covering entertainment, sport, movies and news with Astra providing the vehicle for distribution into the UK. Even before the satellite had launched, News Corp had committed to contracts with upfront payments, which for a start-up venture is almost unheard of.

At a later stage, News Corp began its flirtation with the British DBS project. After a fierce and well documented battle BSB, the nationally awarded DBS operator merged with Sky. Astra benefited immensely as it became, and still is, the satellite home for the BSkyB project and the home for satellite reception in the UK.

91/06/25
Croatia and Slovenia declare independence

91/07/10
Boris Yeltsin first freely elected Russian President

The Foundations are Laid

On my arrival in Luxembourg end of 1985, SES consisted of a single office floor with a dozen employees: two technical experts, an accountant, a PR manager and a few secretaries. I first set out to find a very good commercial and a technical director, with whose help we gradually managed to form a strong and efficient team. Moving later into our home at the Château de Betzdorf was the answer to our increasingly precarious space problems.

In my view, our project could only be successful if a large number of people could receive its signals. In our conversations with the broadcasters, the satellite's reach (and power) were the only arguments that impressed them. It was extremely important for us to build a portfolio of channels which were attractive for the different target countries. This meant reserving capacity on the first satellite even though there were no early commitments in some important markets, such as Germany.

Pierre Meyrat
SES Director General, 1985-1995

91/08/25
Michael Schumacher makes his Formula One debut (Belgian Grand Prix)

91/09/28
Jazz trumpeter Miles Davis dies

The UK helped a lot, of course, since Murdoch knew at a very early stage that he could only be successful with direct reception of his programmes, given that UK Cable did not provide an adequate reach. Some of the German channels were rather interested too, given their dismal view of cable's slow development.

We needed to ascertain availability of inexpensive receivers and dishes, which were easy to operate and install. Hence we opened marketing offices in all the important potential markets, where advice could be given and from where we could support the local manufacturers and retailers in the promotion of suitable receivers – which we then approved as "Astra compatible". Thus these manufacturers and retailers became the sales promoters for Astra, and the reason for which our marketing concept worked so well.

"The launch of Astra 1A meant huge anxieties for everyone on the team. It was the culmination of three years of dedicated and unrelenting work."

Marketing was the decisive factor. A marketing budget for reaching a hundred million people in Europe would have been immensely expensive; hence we had to look for allies. Financially this was the only way to achieve a good marketing result. Clearly the Astra broadcasters had an interest in marketing their services. But given that potential viewers had to install new equipment to receive the signals from the satellites, it was essential to make manufacturers and retailers of hardware strongly motivated allies. The boom actually took place in the shops. Above all, receivers and dishes had to be easy to install and to handle.

Before us, nobody had given marketing much thought. Eutelsat put technical miracles into orbit and said: "whoever wants to can rent them", never bothering about ways of making the service available to consumers. Contrarily, our idea was to START with the consumer. This was Astra's novel approach, whereas thus far satellite operators like Eutelsat and Intelsat had appeared to care only about renting transponders.

Whilst marketing was a demanding, though manageable activity, we were confronted with far bigger challenges in the form of political resistance from a number of important circles. To begin with, the various Post and Telecom organisations, united in Eutelsat, were firmly bent on obstructing all competition. Moreover, two of the major European manufacturers of television sets wanted to introduce (by EU directive) a new, purely European, TV Norm – the so-called D2-Mac – which would have delayed the Astra debut by years, might even have killed our project altogether.

Thus, the major tasks were neither of commercial nor technical nature, but boiled down to sheer political manoeuvring. Fortunately, negotiating bureaucratic obstacles in Great Britain proved to be relatively easy at the time, thanks to the prevailing liberal climate of the Thatcher era.

Our first deal with British Telecom (BT) was a milestone, given that it was, at that time, the first and only privatised telecommunications company – and, what's more – one of the biggest partners in Eutelsat. In Astra, BT perceived a viable and possibly much more flexible alternative to realize its plans for a wider choice of new television channels.

Getting one of the major players of Eutelsat on board was viewed as a first, albeit weak, sign of political breakthrough. Our first customer was Rupert Murdoch's News Corporation with four English language Sky channels. Not only did the deal include highly interesting overall conditions for us; what's more, Murdoch was ready to hire the four transponders with advance payment for ten years. Those were rather exceptional

91/10/27
First free elections held in Poland

91/11/05
Tycoon Robert Maxwell's body found floating

91/12/19
Paul Keating beats Bob Hawke in Australia elections

concessions in favour of a startup company like SES! Before long three more British programme providers followed suit.

For the launch of Astra 1A we had seven British and three Swedish programme providers on board. Had we chosen to peddle the remaining transponders, our service would have turned into patchwork with limited attraction for the TV viewers.

There was but one other language sector that could add to Astra's attraction. Expansion of the German cable network made very slow progress; its construction had started only a few years ago by the Bundespost, the German telecom organisation, whilst the German programme providers were in dire need of a much wider reach. Thus Germany was, apart from Britain, the most promising market for direct reception. The major broadcasters, such as PRO7, RTL and SAT1, were obliged to use the Bundespost's satellites TVSAT and Kopernikus.

After a thorough market analysis, we persuaded them to make use of Astra – in addition – by offering a variable transponder hire rate that was directly linked to their resulting growth of sales. Our board of directors approved this deal with great reluctance only – but fortunately the early loss of revenue was more than compensated for in as little as two years.

Shortly after signing of that contract, ZDF, as the first public sector broadcasting organisation, hired transponder 16, our last one. Dating back to the time of our Swiss pay-TV venture, I had known Leo Kirch, the great German media mogul, for some years. He was not only a major partner in several German private sector channels, but also the most important distributor of motion pictures. When I asked his opinion on the Luxembourg satellite venture, it was instantly obvious that he had detailed knowledge of the project and answered that if they managed to get started, launched and marketed, he would be on board!

It's a fair assumption that the fall of the Berlin Wall did speed up the German broadcasters' decisions in favour of Astra. This was so asserted by the responsible director of ZDF, who has been a personal friend of mine ever since. Though it had no direct bearing on our project (Eastern Germany had not been included in our original market survey), but the windfall of roughly ten million additional potential viewers had advanced our break even by something like twelve months.

Our launch was a spectacular success. With all the transponders sold, the one predominant topic was the need of a second satellite, the Astra 1B. There were a couple of internal challenges to overcome and it took a year to conclude the respective contract. This extra satellite opened up a good deal of new opportunities. Huge success was instantaneous in countries where we were already present with Astra 1A. Although we had generated considerable interest and support from key Spanish broadcasters, we unfortunately knew that Spain would have to wait until 1C and 1D came along.

A camera team in front of the Brandenburg Gate after the fall of the Berlin Wall

"It's a fair assumption that the fall of the Berlin Wall did speed up the German broadcasters' decisions in favour of Astra."

On the other hand, France proved by far the most difficult market. There our intense promotion efforts met with huge formal resistance. Eventually it was our persistence with Canal+ that opened the door to this market.

That we would have to find bigger premises for our growing business was a foregone conclusion, however that the Grand Duke's former residence should become available for the purpose, was an unforeseen stroke of luck. It was a hard fight with the board of directors to get the funding of the necessary renovations approved, but in the end, the Château proved ideally suited for the housing of our team, which had grown to just over a hundred by then. What's more, the ground station that was erected next to the Château, turned into an impressive technical marvel that was, after the first satellite launch, visited by a various heads of state of our target countries, including Margaret Thatcher and Helmut Kohl.

The launch of Astra 1A meant huge anxieties for everyone on the team. It was the culmination of three years of dedicated and unrelenting work. After three years of diligent personal effort and unlimited commitment, all of us were staring at that rocket standing there, a huge 58 metres in height. We went to the launch observation site on the night of the event: the countdown started.

Apprehension, worry and adrenaline rose when the countdown was abandoned; the technicians had to change a key part of the rocket. All that highly explosive fuel had to be drained off, the faulty item replaced and the rocket re-fuelled. The next evening excitement built up fast – and yet, the countdown was interrupted again, a few seconds before takeoff... But the launch did take place half an hour later – to a huge cheer.

Grand Duke Henri and Board Chairman Pierre Werner had especially travelled to Kourou (in French Guiana in South America) to see the launch, and back in Luxembourg, Prime Minister Santer and the whole SES team followed the launch on live TV. Had something gone wrong, the investors would have recovered their money from the insurance. But it's a fair assumption that the whole venture would have been scrapped after a failed launch. An awful lot was riding on that first rocket – much more than meets the eye!

You do not argue with success; on basis of this old principle, all those political obstacles and objections dissipated rather quickly. That D2-Mac norm, so fervently lobbied for by some industrial conglomerates, was put on ice by Brussels in due course. The resistance from telecom enterprises and Eutelsat was eliminated in three steps: First, in Britain, through our agreement with BT in 1988. Then in Germany a year later, when, after lengthy negotiations, the Bundespost and the respective Minister consented to the feeding of Astra programmes into their cable network. Finally, in most other countries we were helped along by a fortunate coincidence: an old friend of mine had taken the Swiss Federal Post and Telecommunications Company to the Supreme Court, fighting their prohibition of direct reception of satellite signals. He initially lost his case but subsequently won it in Strasbourg at the European Court for Human Rights, in May 1990.

Strasbourg's very timely verdict was a major milestone in the history of telecommunications, since it robbed the telecom companies of any basis or right to hinder free reception of signals from satellites. Moreover, this epoch-making verdict arrived at the perfect moment to give a handsome boost to all our promotion efforts for Astra satellites.

> "The whole venture would have been scrapped after a failed launch. An awful lot was riding on that first rocket – much more than meets the eye!"

92/04/29
Three LA policemen acquitted in Rodney King case

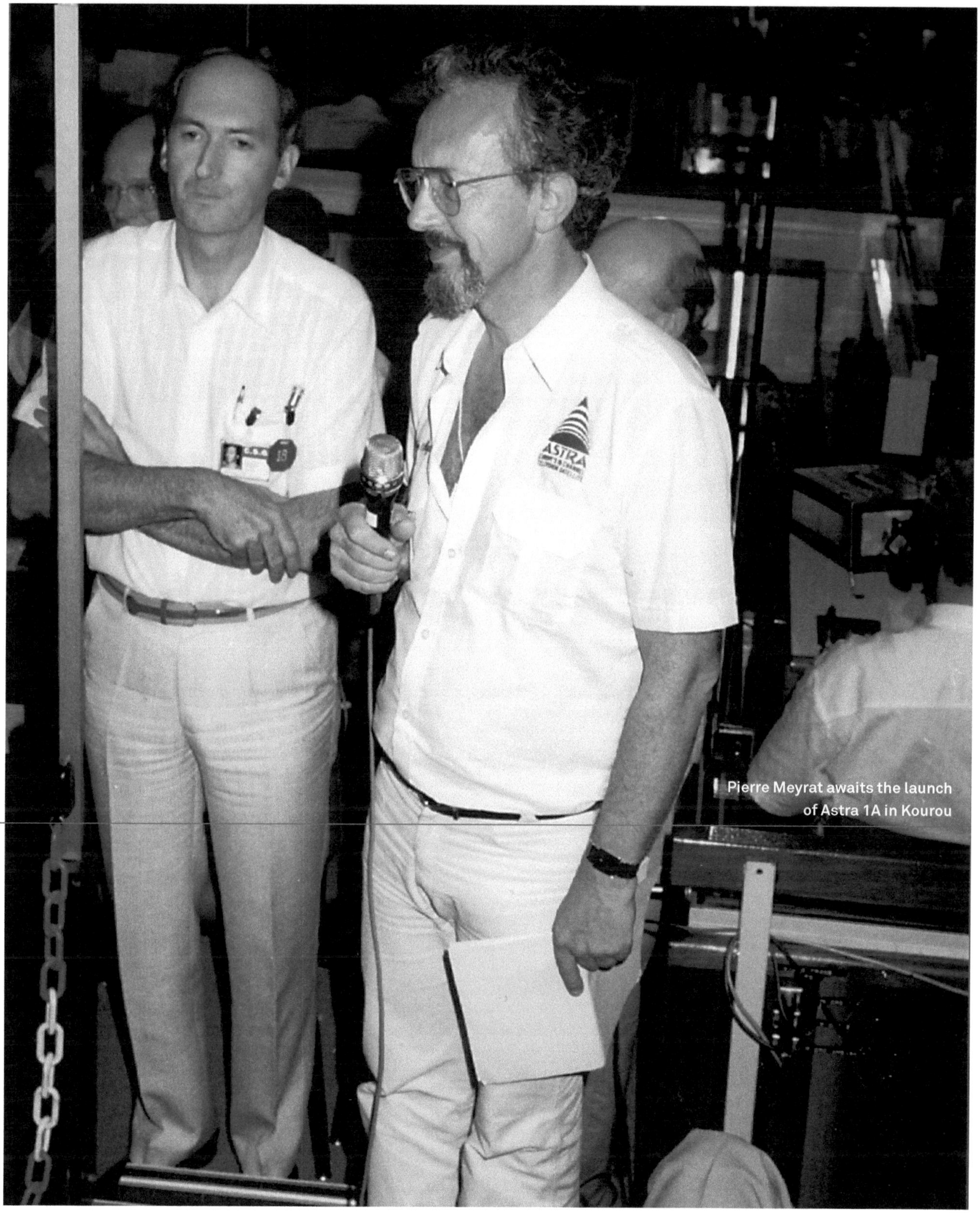

Pierre Meyrat awaits the launch of Astra 1A in Kourou

92/05/23
Mafia bomb kills Italian judge Giovanni Falcone

92/07/01
Bill Clinton nominated by Democratic convention

92/07/25
1992 Summer Olympics open, Barcelona, Spain

Winning the Clients

The biggest challenge for Astra in the early days was that it was a step change in communications and television. So whereas people had their four broadcast channels, all of a sudden we were proposing the choice of 16 channels and then 32 and then 64 and then the hundreds of channels which we now know. With that step change, there came a tiny company, in a tiny country, not one of the major parts of Europe, and we were combating other major competitors and at the time in 1986 there was the small matter of a lack of capital: we needed 350 million euros. We had no frequencies, we had no regulatory environment, we had no clients, we had no satellite, we had no programmers, we had no programmes and we had no reception equipment. Otherwise it was totally perfect!

Jim Styles, Rupert Murdoch, Alan Sugar, Markus Bicknell

Astra was a perfect concept. The team, including Johnny Johnson and Clay Whitehead, had these great visions and we had to just buckle down and actually make them into a business. It was not easy and it took us two to three years to convince the market that there would be a demand. In fact, McKinsey in Germany proved with its research that there was no demand, zero demand for any channels beyond what the

"Astra simply proposed the right thing at the right moment, so 16 channels of television, at a low distribution cost to get to tens of millions of homes, suddenly rang the right bells."

ARD and ZDF were providing at that time! But the more we talked to the major players who were entrepreneurial, and who wanted to combat the status quo of broadcasting, they thought there was an alternative to TF1, ZDF and the BBC. They thought that people would adopt a different view of television. And they found that we were the vehicle that would take them into their homes. We just had to play on that, but it was a long and complex task.

Marcus Bicknell
Astra's first Commercial Director, and now a member of the SES board

92/10/31
Catholic church reinstates Galileo Galilei, after 359 years

92/11/03
Bill Clinton wins US elections, beating Pres.George H W Bush

Marcus Bicknell hangs out with Bill Roedy, head of MTV Networks in Europe then and now

Astra did not address an evolution, no. Astra *WAS* the revolution. There was no television other than terrestrial national broadcasters, well, with the exception that Sky was on a small satellite. In cabled homes there was this glimmer of possibility of multi-channel choice. But I think that it was Astra's presence which actually persuaded the broadcasters, including Canal+ and Murdoch. They saw that Astra now had a mechanism that can actually deliver. And it was not a nationally sponsored DBS service with five channels but coming from these entrepreneurial, funny guys in the middle of Europe.

Astra simply proposed the right thing at the right moment, so 16 channels of television, at a low distribution cost to get to tens of millions of homes, suddenly rang the right bells with those entrepreneurial broadcasters. One of the greatest achievements was listening to those broadcasters, discovering what was it that they needed, how many channels, over what area, how could they get the equipment down to a certain price. In determining what they needed, we realised that the proposal of the 16-channel satellite was the right one and history proved us right.

However, there was huge opposition to Astra. The major countries around Europe: Germany had its Medienanstalt, the media authorities, and France had its own regulatory commissions. I remember that the [French] Minister of Communications in 1988 declared that he would send up some spoiling signal to prevent us being received. The opposition was very strong; we felt it and it was painful.

92/11/24
American forces quit Philippines, after almost 100 years

92/12/09
Prince Charles and Princess Diana announce separation

Our primary target was Sky Television and Rupert Murdoch himself was calling the shots. The decision was not something he delegated. And we had competition, because he was being solicited by BSB, the five-channel British satellite and by others. We also forget [the Atlantic project] which was similar to our own and based in Ireland. And we had to be pretty clever. John Tydeman was probably the most significant SES employee in pulling Murdoch in. We met Murdoch several times, as well as his team from America and from Britain and we suddenly got to the stage where we realised that we could have a multi-channel deal. And indeed on June 8, 1988, he announced the signature of those first four channels. One cheque for 80 million pounds, then worth 180 million euros, for the use of four transponders for 10 years and from then on Astra was made. That was the turning point.

I think Astra's biggest achievement is having stuck to its guns. There was the straightforward idea: 16 channels, medium-power, a 90-cm dish; and it all worked. Ever since then, SES has been a company totally dedicated to satellite distribution with the best possible reach, serving the best possible broadcasters. I think the Luxembourg government, the local environment, everything that's serious about what happens in Luxembourg played a very large part in all of it. We haven't had to chop and change. We were right at the beginning, and we're still right today. And it's very interesting being back on the board of SES 25 years later, having done lots of other things. This is a serious company; they really do this well and because it was good at the beginning, it seems to be good now and nobody wanted to change that fundamental goodness, those values, that make SES what it is.

I also think that one of the strengths of SES was to include the Luxembourg state banks and their ethics, their values, which are about being solid, being true, being honest, doing what you do best and doing it really well. And I can remember things which signified the early days of SES and the Luxembourgish influence. My father had impressed one thing on me: if you are going to do something, you might as well do it well. When we modernised the Château, I was part of the team that conceived how the central atrium would look. And it was done as a high quality task. We spent a lot of money. But that marble 25 years later, it is as shiny as the day we put it in. It is the most extraordinary testament to doing things well.

"When we realised that we could put more satellites in the same position, that we could go to multiple channels on the same dish, suddenly we realised that this could be a bigger dream than we expected."

In 1986, you could see what it would become. Was it just a 16-channel satellite? On the first day, it was difficult to tell. But I did have this conviction about television without frontiers, about television channels being available throughout the whole of a continent, or continents where there would be no borders. When we realised that we could put more satellites in the same position, that we could go to multiple channels on the same dish, suddenly we realised that, no, this could be a bigger dream than we expected, so let's get the basics right and then we can build on that very quickly. It was very exciting. I think we all had the conviction that it could be a world beater.

1993

93/01/01
EuroNews is launched in Europe

93/01
There are just 50 World Wide Web servers in existence

When Canal+ turned to Astra

Marc Tessier
Director Canal+ International, 1989-1995
President France Télévision, 1999-2005

Canal+, which went on air in 1984 as a single terrestrial channel throughout France, was looking to expand with more niche services. Mark Tessier explains: "We believed that our premium pay channel didn't saturate the market. But we also thought there was only a need for a few more channels!"

Initially Canal+ went onto the Franco/German TDF1/2 satellite. "It was a very powerful satellite using a new technology invented by Thomson and a new transmission standard, using D2-Mac. It could handle only a few channels. The frequencies had been allocated to each country by an international agreement and, in France, the ancestor of the Conseil Supérieur de l'Audiovisuel, organised a tender. Canal+ was chosen to carry the project. At the beginning we had a partnership with the Bertelsmann group for this task. At the same time the British had chosen the same type of satellite and it was supported by the ITV group. But there was an independent non-aligned operator, the Murdoch group using a Luxembourg satellite, Astra."

CANAL+

Tessier says that he and Canal+'s president at the time, André Rousselet, supported TDF1/2 because it used a French technology, had a more powerful signal and needed a smaller dish. "And it was a project supported by the French authorities. But it never worked. The launch of the first satellite failed and the second satellite lost a few channels. Its technology was very sophisticated but too fragile."

'Les Guignols', a breakthrough satirical show for Canal+, and still on air

Despite this setback, Canal+ still avoided Astra. "France had its TV and cinema [support] system, based on the notion of 'cultural exception'. French channels had to pay taxes and had investment obligations to support French programming. We all believed that if a channel was transmitted from another country it would not respect our rules. Today we know it is not the case; very few channels targeting the French market are based in other countries. We chose a France Télécom satellite. We planned to broadcast around 20 low-cost pay channels and services in analogue. France Télécom had a communications satellite available that we could use right away. And, less than two years after our launch, we already had around 200,000 subscribers. But we realised that a new technology, digital broadcasting, would spread very fast and that we could transmit many more channels and services than analogue for much the same cost. It would enable us to increase regularly our services at a low cost. In the beginning we worked with Rupert Murdoch's group with which we had a partnership. We had no reason to increase our analogue subscribers."

Asked why Canal+ did not transmit digitally on its France Télécom satellite, Tessier says: "We planned to broadcast as soon as possible initially 30 channels, then 50, and then even more. And we realised that the television industry could not expand by relying on proprietary, expensive and controlled technologies but it needed open technologies that would become cheaper and cheaper (for us and consumers). France Télécom was not hostile to digital but simply didn't

93/03/01
Siege begins at Branch Davidian cult compound, in Texas

93/03/22
Intel Corp ships its first Pentium chips

"We realised that the television industry could not expand by relying on proprietary, expensive and controlled technologies but it needed open technologies."

have enough satellite capacity available to offer us more than 15 digital channels. And it had also been approached by [commercial network] TF1 which asked to share its digital capacity to launch its own satellite service, a competitor to Canalsat. Thus France Télécom offered us only a capacity for 15 channels, which was nonsense as we already had 20 channels in analogue."

So Canal+ turned to Astra. "One of the strengths of Astra was that it had anticipated all of the needs of television. When I first went to France Télécom to reserve satellite capacity, they gave me a one-page sheet of paper to sign. In this document France Télécom said would take care of everything. It guaranteed that its satellite would work, that it never had had a problem! They didn't go after clients and never negotiated. And, me, as any private company, I needed precise guaranties, listing all the possible problems and telling what would happen in each case, so I could take out suitable insurance. But the president of France Télécom didn't seem even to understand my question. It was the sort of logic that came from public services at the time. Astra immediately gave written answers to each of those questions.

Its management understood the logic of a private client. I remember a conference in London on pay TV. There was the president of the public satellite operator, BSB and the Murdoch group, Sky TV that was already on Astra. The first one had arrived in a Jaguar, the second in a taxi. When the first was asked the price of its decoders he asked his technical manager to answer while the second said that it was the key point, that he had negotiated a 40% rebate from his suppliers and he gave the exact price. Two different worlds."

Tessier admits that signing with Astra created something of a shock in France. "I remember, it was in October 1994. Pierre Lescure, then president of Canal+ and I called Pierre Meyrat at SES. He arrived alone, with a letter of intentions of a few pages, but very detailed, and we all signed it. We asked him about the press communiqué. He said that he knew this was a very touchy question. We explained that as a public company we had to publish a communiqué. When he left he said "Bonne chance". And it is true, as soon as the agreement was known, we received a lot of phone calls. The fuss soon died down."

Cable plays an important role in many European countries. Mr. Tessier explains the view from France: "France is a non-typical country. It is the country where cable failed. As in Great Britain. The reason for this failure is not that French or British viewers didn't want a hundred channels. It is that there was a succession of French cable plans, with new technologies, each one incompatible with the other. Cable was thus always too expensive to be attractive. The French market was open for satellite. Besides, in France, the competition between two DTH operators forced them to over-invest in marketing and reduce their prices."

Marc Tessier says he admires Astra for having made the right satellite decisions. "Also, the choice of a technology that could be received all over Europe and thus benefit from a mass market. A small, but very competent staff that was quick to respond. And the Murdoch team helped them sharpen their methods. What is striking is that, in Europe, the penetration of the different television broadcasting technologies varies from one country to the other. It proves that much depends on the operators. In France and the UK, the success of satellite is mainly due to the investments of Canal+ and the Murdoch group. In Germany cable operators were more dynamic. Today France represents 25% of ADSL's TV reception around the world. It is due to the dynamism of the three triple-play operators and their marketing policies. Thus, anyone who says he knows which technology will prevail ten years from now is presumptuous."

93/05/12
Astra 1C launched (French Guiana), adds 18 channels

93/04/30
Virgin Radio launches in UK

93/05/27
Car bomb at the Uffizi Gallery, Florence, kills 5

Astra: The name is very popular!

Google the word 'Astra' and you'll come up with some 70 million entries. About 69 million of them concern the sale of assorted Astra vehicles especially in the UK and North America! Besides our favourite satellite operator, readers might be interested in these other Astra organisations, brands and descriptions.

Astra is from the Greek, and means 'star'. The Astra Planeta were the gods of the five wandering stars (or planets).

The Astra brand is used in the UK and Ireland by the General Motors car subsidiary Vauxhall. You will also see Astra models as part of the US and Canadian GM/Saturn range of cars and, of course, from Opel in Germany.

Astra is also the name given to a car in the USA, and in particular to a very pretty 1952 design which used an Oldsmobile Rocket 88 V8 engine and chassis, with a special custom body designed by noted stylist George Barris.

Astra is also: Association in Scotland to Research into Astronautics.

Citadel Astra (Automated Spectrophotometric Telescope Research Associates.

A CCD-based spectrophotometer for a new automated 0.5-metre telescope at the Fairborn Observatory, Washington Camp, Arizona, USA, built at altitude 1,800 metres should see first light in 2009-2010.

The Australian Subscription Television and Radio Association (Astra) is the main industry body for subscription television.

Not to be outdone, Astra is also the Australian Strategic Air Traffic Management Group

---

Astra Film, based in Romania, is a centre for documentary film in Romania running a well-established international film festival of documentary film.

Or how about the American Specialty Toy Retailing Association (Astra) which was founded in 1992 on a simple premise: to bring members of the independent toy industry together to help each other succeed.

Astra is also the name of a 1985 music album by rock band Asia. A more recent Astra is the name of a Californian pop group. Their latest album, 'The Weirding', was released in 2009.

Astra is the abbreviation for the South Carolina-based Advanced Security Technology Research Alliance "where our mission is to create a united voice and message on cyber security capabilities".

Astrophysics and Space Sciences Transactions (Astra) is an international scientific journal dedicated to the publication and public discussion of high quality original research on all fields of Astrophysics and Space Sciences and related technology.

Astra is a Bulgarian non-governmental organisation dedicated to eradication of all forms of trafficking of human beings, especially of women and children.

The Applied Science and Technology Reservoir Assessment (Astra) Initiative at the Kansas Biological Survey (KBS) was created in 2006 in response to the need for critical information concerning the status and conditions of Kansas reservoirs.

Astra is an indie music club in Berlin.

Astra is also a female superhero, or at least a super-villain, in Marvel Comics. She first appeared in Uncanny X-Men, issue 107 in 1977.

Astra is also the Sanskrit word for missile, and in the long war between the Pandavas and Kurus that's related in the epic Indian Mahabharata there were astra weapons, mantra astras (weapons made more powerful by a spell or mantra), divya-astras (weapons received from higher beings).

Astra is also the name of an important brand of fireworks (feu d'artifice) in the UK.

AstraZeneka is a giant pharmaceuticals company.

Astra is also the name given to a sub-set of Adobe's Flash technology.

Astra International is an Indonesian conglomerate.

Astra, the Advanced Sysinfo Tool and Reporting Assistant, keeps a watchful eye on computer motherboards.

Astra Games Ltd, based in the UK, makes gaming machines (the sort found in bars and clubs).

Astra is a self-drive car hire company in Cyprus.

Astra is the brand given to a range professional coffee machines.

AdAstra is the name of the National Space Society's magazine.

The 3*** Hotel Astra in Prague, Czech Republic is only one of a large number of Astra-named hotels.

Astra Excursions tailor-make guided walking holidays in Greece.

Astra was also the name of a pistol, which once belonged to Nicaraguan dictator Anastasio Somosa.

06

Europe's Dish and Box Wars

06
Europe's Dish and Box Wars

Chris Forrester

It is difficult today to look back 25 years and remember what a shock satellite television was to the established status quo across Europe. Public broadcasting, much of it good, had an entrenched position, but emerging commercial networks were already shaking up the industry. But satellite would provide the breakthrough, the electric shock to the system, which was badly needed.

In television terms, Europe, according to the advertising industry, comprises 33 countries and 35 sales markets (the difference being the linguistic splits in Belgium and Switzerland). Even though some pan-European satellite channels attempted to break through these national and linguistic barriers some 30 years ago (notably the original Sky channel in 1982, 'SuperChannel' which launched in February 1987 and MTV which launched in August 1987), Europe still remains a fragmented television market.

Even though the official line from the European Community is to support and encourage the concept of "Television Without Frontiers", the reality is that European harmonisation of television services simply has not happened. There is some limited cross-border traffic in complete channels (French channels in Wallonia, Dutch channels in Flanders, BBC services in Ireland and the Netherlands), but these are the exception and generally restricted to linguistic groupings.

The potential market for direct-to-home satellite reception was absolutely minuscule – no more than 113,880 installations across Europe.

93/06/20
Japan, 7.5 earthquake kills 385

93/07/27
NBC TV pays $456m to cover 1996 Olympics

93/08/09
King Albert II of Belgium is crowned

C. Forrester (ed.), *High Above: The Untold story of Astra, Europe's leading satellite company*,
DOI 10.1007/978-3-642-12009-1_6, © Springer-Verlag Berlin Heidelberg 2011

Some of the early analogue satellite channels were:

Name	Original Ownership/notes	Launched
Sky Channel	Brian Haynes (backed by Guinness Mahon, etc.)	Apr 1982
Screensport	W H Smith, ABC (ESPN)	1983
Children's Channel	BT, Thames, DC Thompson, Central TV, EMI	1984
Teleclub	Rediffusion AG 60%, Beta Taurus (Kirch) 40%	1984
TV5	A "best of French broadcasting" channel	1984
Premiere	EMI, Goldcrest, Fox, Columbia, HBO	Sept 1984
Sat.1	PKS 40%, Axel Springer 15%, APF 15%	1985
CNN	Turner Broadcasting	Sept 1985
Lifestyle	H Smith, TVS and Yorkshire TV	Oct 1985
SuperChannel*	Granada, Virgin, Yorkshire TV	Feb 1987
Satellite Info. Serv (SIS)	Ladbroke, Mecca, Coral, Grand Met, Wm Hill	May 1987
MTV	51% Maxwell, 25% Viacom, 24% BT	Aug 1987
TV3 ScanSat	Kinnevik 96%, Nora (Norway) 4%	Dec 1987
The Movie Channel	British Satellite Broadcasting (BSB)	Mar 1990
NOW	BSB	Mar 1990
Galaxy	BSB	Mar 1990
Power Station	BSB/Virgin	Mar 1990

***SuperChannel had started out life as Music Box**

While the United States had an estimated 36 million TV households by February 1955, the whole of Europe possessed only some 4.8 million TV sets at that time.

There's one other important similarity with viewers in North America. Europeans are turning away from the established public 'free-to-air' networks. Just as in the United States with HBO, CNN and the other broadcasters who sought to reach their viewers exclusively by cable, various broadcasting groups attempted a similar approach in Europe. The early operators tried to imitate the American broadcasting model, intending their signals to reach cable distributors, deliberately making their channels near-copies of commercial television services, although heavily (indeed, almost exclusively) dependent on inexpensive, imported programming.

Besides 'Sky Channel', TEN (The Entertainment Network) launched in March 1984. It went bust a year later, being re-launched as Robert Maxwell-owned Mirrorvision. Mirrorvision merged with (British-based) Premiere in April 1986. Premiere closed in 1989. Sky Channel, despite News Corporation's involvement, struggled badly, as did 'SuperChannel'. These pan-European services, lured by the concept of broadcasting a common entertainment channel to a growing number of well-heeled viewers, found some enthusiastic support from younger viewers but the numbers were never enough to attract adequate advertising support. These early satellite broadcasts were made from low-power orbiting satellites, effectively ruling out their reception 'direct to home'. Astra changed that problem.

Sky Channel was Europe's first satellite TV channel, launched in April 1982. In 1983, News International bought 65% of its shares, and by 1988 owned 82%. In 1987 Sky increased its capitalisation by raising 22.63 million pounds, but its losses to then (and subsequently) were substantial, amounting by the end of its 1987-88 financial year to almost 39 million pounds. Even early launches of popular thematic channels like MTV had tough times, and many forget that Viacom's first partner in Europe for MTV was Robert Maxwell who owned a 51% share of MTV Europe, with another 24% owned by British Telecom. Cabled homes (less than 15 million over the whole of Europe in 1988) were still insufficient to provide a viable market for international services.

The market needed something else to spur it to action, and that happened on December 11, 1988 when Astra 1A blasted off from Kourou in French Guiana. From it, broadcasters supplied a mix of English and German analogue programming designed for reception on small (typically 60 cm) roof-mounted dish antennas.

However, it is worth remembering how very brave this move was. That year (1988), the potential market for direct-to-home satellite reception was absolutely minuscule – no more than 113,880 installations across Europe, according to 'Cable & Satellite Europe' magazine. But the market seemed ready to expand, a fact noted by many potential broadcasters across Europe.

Satellite would provide the breakthrough, the electric shock to the system, which was badly needed.

The Emergence of BSkyB

Britain had been allocated five direct broadcasting by satellite frequencies as early as 1977, but it took until December 1986 for the then Independent Broadcasting Authority to select British Satellite Broadcasting as the operator. BSB, which achieved fame and then notoriety with its 'Squarial' dishes, had a troubled time of it. Promised starting dates were broken. Technical problems beset the project and financing cash was being spent at a spectacular rate with accusations of profligacy at board level. BSB eventually commenced transmissions in April 1990, more than a year after Rupert Murdoch had launched his Sky service.

On June 8, 1988, what has been described as the 'rumpled' figure of Rupert Murdoch announced to a press gathering that News International would launch four 'Sky' channels broadcasting from an Astra craft by February the following year, less than nine months away. "We are seeing the dawn of a new age of freedom for the viewer," said Murdoch.

What followed was probably the bloodiest period in British broadcasting, or perhaps broadcasting anywhere. Two companies, Sky, backed by News International and British Satellite Broadcasting, backed by more than 1 billion pounds of private finance, were locked in a battle for the eyes, subscription cash and loyalty of British viewers. BSB used a new broadcasting standard, the D-Mac system (see 'Mac´s Less Than Golden Thread' section later in this chapter), needing specially-developed receiver/decoder boxes, although using fairly high-power satellites, receiving dishes could be small, no more than about 40 cm across.

Astra 1A went 'live' on February 5, 1989. Although designed as a telecommunications craft, rather than a broadcasting satellite, it was the vehicle that enabled many more viewers to access satellite TV with low-cost, rudimentary receiver boxes and small satellite dishes. Up to this point, viewers had needed dishes of at least 90 cm diameter (and ideally 1.2 metres) to obtain a decent screen image. Murdoch was first to market, with a low-cost receiver box developed by Alan Sugar's Amstrad company. His February 1989 launch was not without problems, with a promised Disney channel just one of the stumbling blocks. But well within a year of launch, Sky was reporting operating losses of 95 million pounds on top of start-up costs of 121 million pounds (News International/ Sky, June 1990). These were bad, but BSB's losses were worse. By October 1990 the two companies had started merger talks, which were completed in November. According to

1994

94/01/06
Skating star Nancy Kerrigan attacked

94/01/17
Northridge earthquake, 6.7, hits Los Angeles. Kills 72

MTV was an instant hit, growing to dozens of channels over Europe

published estimates, by the time BSB and Sky had merged (to become British Sky Broadcasting) they had spent a total of 1.25 billion pounds creating a direct-to-home market in Britain.

Murdoch and his low-cost service, transmitting its channels on the Astra telecommunications satellite, had won the day, even though losses would continue. Indeed, two reports which circulated in 1988 were not that encouraging of satellite's prospects as a broadcasting medium. Logica estimated a UK dish population by 1994 of 600,000. Even more depressing was a CIT document predicting Europe-wide dish ownership of just 1 million by 1996. There were many similar doom-sayers.

If the 1988 predictions of barely 1 million dish installations across Europe by 1996 were compared to today's real position, then by any measure, the enthusiastic adoption of satellite broadcasting has been a huge success.

Astra's own 1987 predictions of its 1996 position have also been widely exceeded, as revealed recently by Marcus Bicknell, Astra's first Commercial Manager. He admitted that some people thought his forecasts optimistic, but they were based on the ownership of VCRs in any given market. On that basis, Astra predicted that in 1996 there would be 20.2 million DTH and SMATV viewers across Europe (some 17% of European TV homes). The actual 1996 figure was 22.97 million, as Bicknell stated, not only ahead of Astra's

94/02/12
Munch's painting 'The Scream' stolen in Oslo

94/03/12
Church of England ordains first female priests

own estimate but reasonably accurate as to the actual market position at the time. By mid-1999 the figure stood at more than 33 million homes.

However, Bicknell's forecast of European cable connections has proved to be somewhat underestimated. In 1987, Astra suggested total cable viewing of 23 million homes in 1996. The reality was near double, at 44 million homes in 1996 and 50 million at mid-year 1999. Astra's predicted combined total (DTH and Cable) for 1996 was 43 million homes, but the actual numbers were 67 million and now stand at more than 125 million homes. Bicknell recently suggested that instead of basing his forecast calculations on VCR ownership, a better guide might have been total television ownership itself. No matter. The numbers were all positive, and well ahead of the initial estimates. A massive industry was being created.

But there were threatening clouds on the horizon. Just about each European country had its own plans for satellite broadcasting – and tiny Luxembourg did not figure in those ambitions. The UK broadcasting regulator in 1984 wrote formally to Richard Dunn of Thames Television (and an enthusiastic supporter of the Astra concept) telling him that SES was an organisation of "doubtful legality". Dunn promptly sent a copy of the letter to Luxembourg's Ambassador to London, suggesting he take the issue up with the UK Foreign Ministry. Thames, in early 1986 bought a 5% stake in SES, doubling its investment a few months later.

Britain's antipathy to Astra was shared across Europe. It seems that each of the major European nations feared what one described as a "foreign cultural invasion". The Brits feared American programming would damage the position of both the BBC and ITV. The French were keen to protect their own programme-making from any outside influences. Germany had its own raft of protections to take care of. Each of these markets had its own satellites in the sky.

European Satellite Dish Installations - 1988

Austria	6000
Denmark	8000
Finland	3600
France	10,000
Greece	1500
Ireland	2000
Italy	2000
Luxembourg	80
Netherlands	2000
Norway	3700
Portugal	2000
Spain	19,000
Sweden	17,500
Switzerland	1000
W. Germany	19,500
UK	18,000
Total	113,880

Source: Cable & Satellite Europe, Oct 1988

Britain's antipathy to Astra was shared across Europe. It seems that each of the major European nations feared what one described as a "foreign cultural invasion".

94/04
Yahoo is created

94/04/27
South Africa holds its first multiracial elections

Mac's Less than Golden Thread

Julian Clover

The road to high-definition television has been very long, very challenging and not without the creation of more than a few dead ends. The biggest problem came about during the 1980s and early 1990s, and revolved around the Mac transmission standard. Mac (Multiplexed Analogue Component) television was designed with the best of intentions as a sort of half-way house between standard analogue (which in Europe meant PAL or Secam transmissions) and full digital TV.

Mac was developed by the UK's Independent Broadcasting Authority (IBA) and formally adopted as the transmission method on the UK's DBS (Direct Broadcast Satellite) in 1982. A year later it was taken on by the European Broadcasting Union as the future TV standard for satellite broadcasting in Europe.

In many respects the EBU was working hand in hand with the TV set community, as well as with its public broadcasting members. Their concept was straightforward: Mac would be used as a launch pad for a European widescreen television system. The British quickly changed their specification, which made life extremely difficult for the consumer electronics industry because it needed two separate demodulators to be included in the receiver. It was also incompatible with cable transmission. The French came forward with their solution: D2-Mac.

94/04/29
First ever free national elections in South Africa

94/05/06
Channel Tunnel opens after 7 years of building

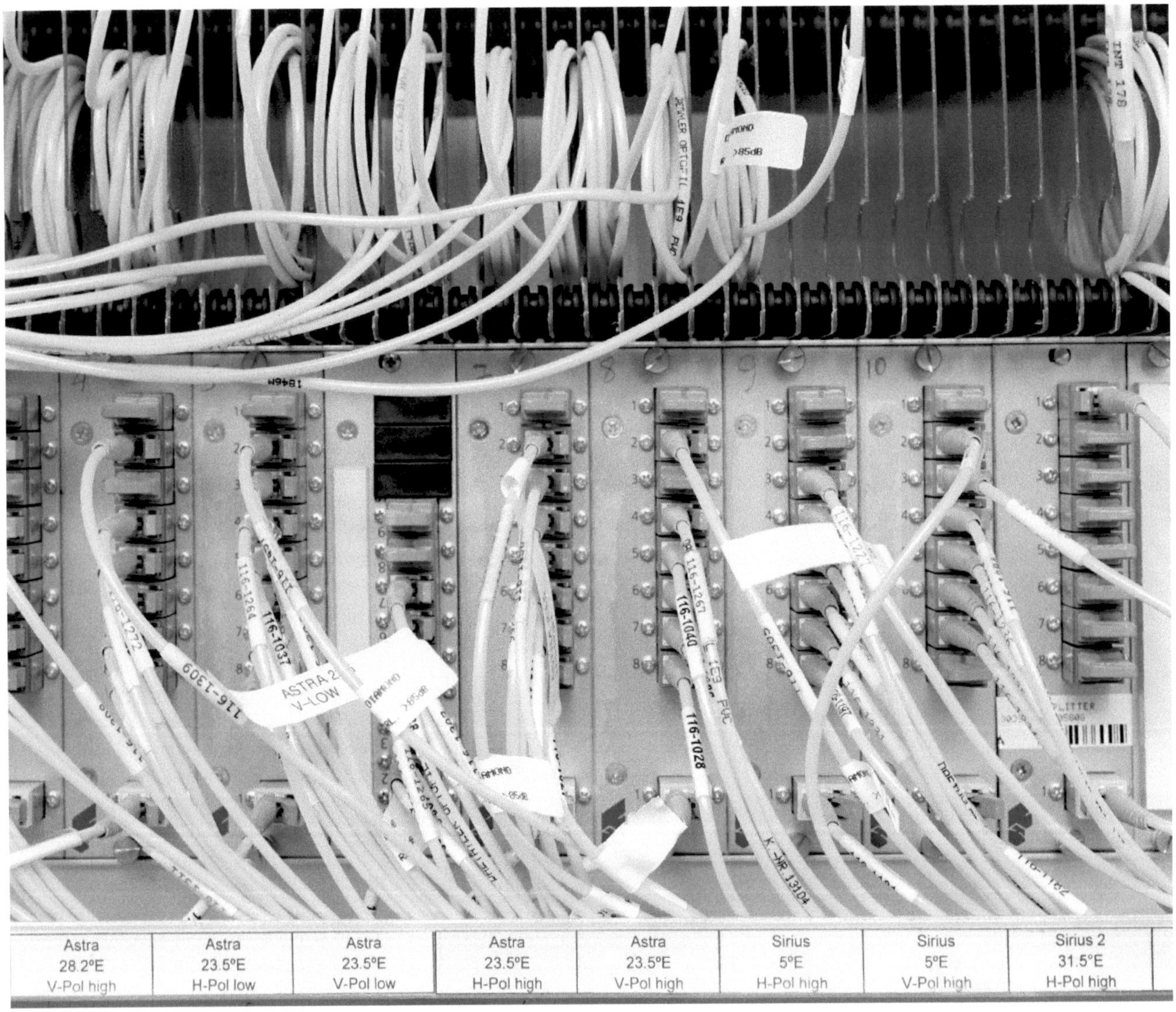

94/06/18
OJ Simpson arrested for murder of wife and Ron Goldman

94/06/24
First French "all news" channel (LCI) begins

Across the other side of the skies, a series of satellites was emerging under the 1977 WARC plan. The frequency planning conference designed a plan that would only offer around five channels per country.

Germany joined France while Norway led the Scandinavians in joining Britain. It quickly became a technical nightmare, with nobody keen to give way. In 1986, the Euro-Mac consortium was formed by Europe's leading set manufacturers in an attempt to kick some life into the mess. The same year, a European Mac Directive imposed the use of Mac on the various Direct Broadcast Satellite (DBS) systems that were being established by individual member states. A meeting in July 1987 tried to convince the British to opt for D2-Mac, but British Satellite Broadcasting (BSB) was already moving ahead with plans to use the agreed D-Mac system, signing up technology giant ITT-Intermetall to produce dedicated chipsets that only added to the delay in getting BSB onto the air.

Alongside this near-nightmare there were continuing pressures on broadcasters to create new techniques to get people into the retail stores and buy new TVs. Pal+, a widescreen technology, was backed by plenty of TV set manufacturers (and in those days there *were* plenty of European set manufacturers), and Pal+ helped to shift some TV sets out of the stores. Of course, hardly anyone – other than Hollywood - was making programmes in this new widescreen format, but Grundig, Philips, Schneider and Nokia all produced sets that were Pal+ compatible.

Astra's use of FSS (Fixed Satellite Services) frequencies rather than those in the BSS (Broadcast Satellite Services) band meant that Astra wasn't obliged to adopt the Mac technology, but initially the Luxembourg operator was supportive, or at least technically agnostic as far as its clients were concerned. Even Rupert Murdoch considered the system for the launch of Sky Television. Ultimately, Astra emerged as an opponent of D2-Mac, lobbying against the system and trying to encourage discussion around its viewpoint. "The fact that some major European manufacturers of television sets wanted to introduce, by EU directive, a new, purely European, TV norm would have delayed the Astra debut by years, might even have killed our project altogether," says Pierre Meyrat, former SES Director General. Murdoch, like Astra, wisely, didn't wait, thereby sidestepping any problems with expensive chipsets and with Alan Sugar at his side, unveiled the Amstrad receiver and dish that would help launch Sky in February 1989 with the ever-reliable, and well-understood PAL transmission system.

The beginning of the end for Mac came in November 1990 when the losses pouring out of Osterley (Sky) and Marco Polo House (BSB) forced a merger between the two and the forming of BSkyB. The Murdoch business had gained the upper hand and the UK had lost its Mac champion. The consumer electronics business went into a state of panic and headed to the European Commission in an attempt to get a second directive.

A series of proposals was put forward leading to a vote in the European parliament in November 1991. At the centre of events was the Italian politician Filippo Pandolfi, the European Commissioner with a portfolio for Research and Development. This was the time of the French political Jacques Delors and memorable Sun tabloid headlines "Up Yours Delors".

Pandolfi's proposals had come in for a mauling at the hands of the economic and monetary affairs and industrial affairs committee (EMAC). But Pandolfi was prepared to compromise, accepting that other high-definition broadcasts should be accepted alongside HD Mac and that D2-Mac chips should only be mandated for 16:9, rather than all television displays and satellite receivers. However, there was a typical Euro-fudge on which channels should be required to broadcast in the D2-Mac format, or whether simulcasts would be allowed, and what would happen to those broadcasters already on the air.

94/07/16
Anna Nicole Smith (26) weds billionaire J Howard Marshall II (89)

94/08/10
Last British troops leave Hong Kong

Significantly, and in order to get the agreement of all sides, the European Parliament allowed a Letter of Intent to supersede a previously drafted Memorandum of Understanding. This and a helpful misunderstanding that omitted the word "only" from the passage "HD Mac may be used" completely changed the outcome, opening up loopholes, and arguably plunging the knife deeper into a corpse already beginning to go cold.

The November 1991 European Directive was designed to remain in force until 1997, by which time Mac had been consigned to distant memories as operators prepared to launch their true digital services. In addition to the rules regarding the inclusion of DVB chips, it was decided that all conditional access systems should be compatible with D2 Mac and authorised by a European standards body.

Key to the satellite world was that Mac should be compulsory for all WARC-77 DBS services, except those where the coverage of the satellite also included Latin America, instantly providing a 'get out of jail free' card to the Spanish Hispasat operator. The Directive, helpfully, did not cover Astra's FSS transmissions, to where broadcasters were increasingly migrating.

Across the other side of the skies, a series of satellites was emerging under the 1977 WARC plan. The frequency planning conference designed a plan that would only offer around five channels per country – not even enough for MTV when compared to today's plethora of stations – each with a beam that would largely reflect the country to which it was broadcast.

The DBS satellites that emerged required their broadcasters to use one of the Mac variants; Tele-X, which was located at the five degrees East position now occupied by the SES-Sirius fleet, was a share between Norway and Sweden. The Norwegian public broadcaster NRK went for D-Mac, but the Swedish commercial channel somehow managed to broadcast in PAL. Neither exactly extended consumer choice for anyone with a terrestrial antenna.

Under the so-called Mac Directive, "Meaningful Funding", was to be given towards D2-Mac transmissions of encrypted services from January 1994 onwards, and for non-encrypted services starting in 1996. An initial 850 million euros were set aside for the support of D2-Mac and HD Mac in the following five years, though the figure was later watered down. Britain's representative, the Conservative parliamentarian Edward Leigh, described the proposals as "totally unacceptable". The UK was not alone in its criticism, with Germany, Italy, Spain, Denmark, Greece, Ireland and Portugal all expressing reservations.

Astra did have some stations that were happy to transmit in D2-Mac. TV3 initially thought of using B-Mac before moving to the D2-Mac system ahead of their Astra launch. Sister channel TV1000 and Scansat, ironically broadcast from London, used D2-Mac until their departure from 19.2 degrees East in 1996. They continued to use the format until their analogue shutdown.

The newly re-branded France 2 began 16:9 transmissions in D2-Mac in 1992, simulcasting its transmissions between Télécom 2A and TDF, ahead of plans to leave TDF altogether the following year. It was estimated that of the 45,000 dishes pointed towards TDF, 95 % belonged to Canal+ subscribers, who for the most part could receive France 2 terrestrially. Canal+ was looking to supplement its own terrestrial delivery, but its mix of satellite positions was costly in those single–analogue–channel–per–transponder days and by 1994 the pay-TV channel was issuing letters to the 50,000 subscribers to its D2-Mac service, inviting them to move to the Télécom 2A satellite.

The French had seen Télécom 2A as the Gallic answer to Astra, but the craft was beset with political problems from the word go, Canal+ having initially refused to participate in the satellite at all if it could not use the Secam transmission standard used by France in its terrestrial broadcasts. The trade-off permitted the Canal Satellite bouquet to broadcast in Secam in return for the launch of two, eventually three, D2-Mac

94/08/13
US Major League Baseball players go on strike

94/09/28
Car ferry MS Estonia sinks in Baltic Sea, kills 852

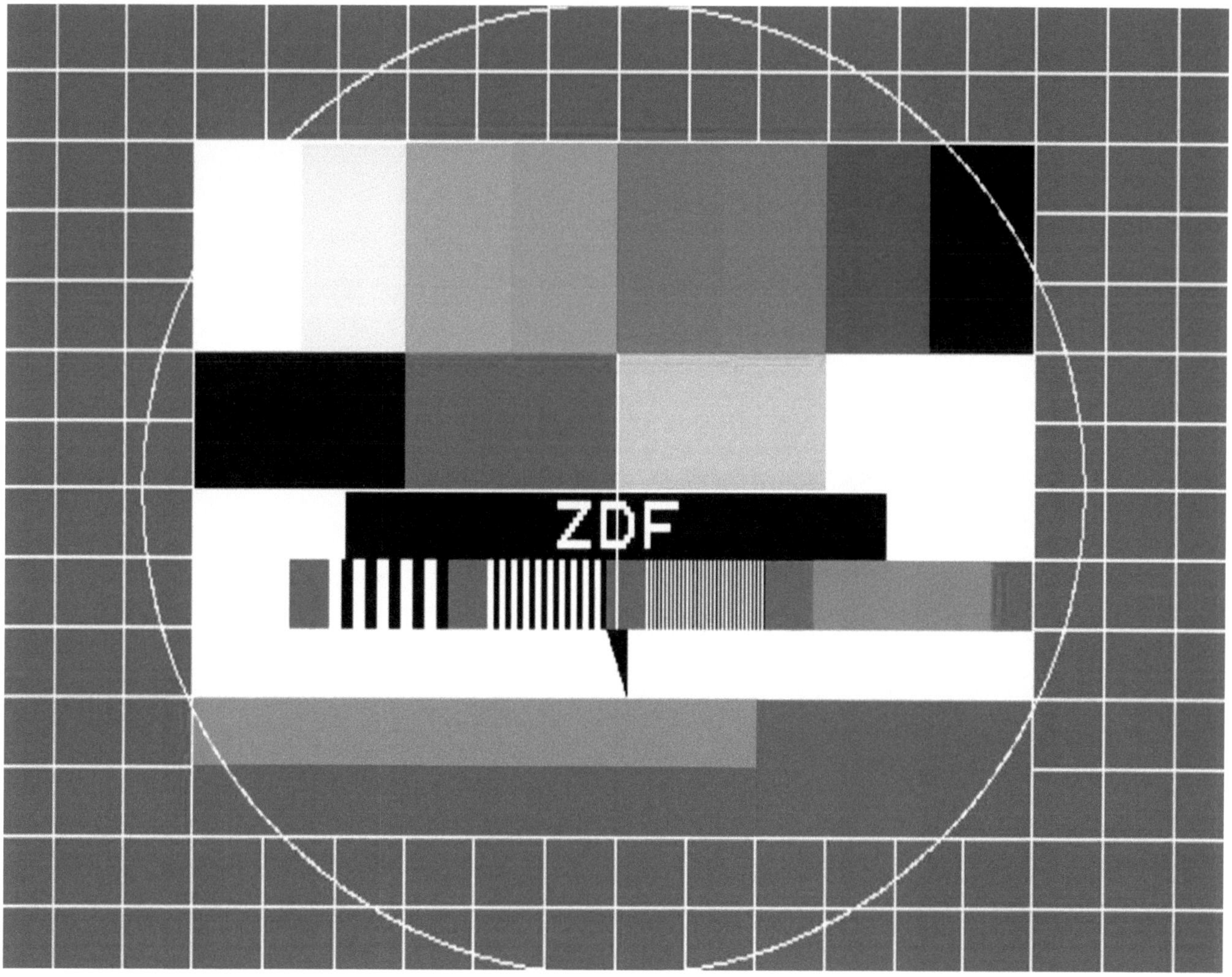

channels. It was at this point that Astra first emerged as the eventual destination of French direct-to-home satellite, though the country has never fully committed to a single slot, the Eutelsat Hot Bird and the positions occupied by the old Télécom fleet and 5 degrees and 8 degrees West continue to be used to a lesser degree.

The worst problems were experienced by TV Sat. The first of the two German satellites was launched on November 21, 1987. One of its two solar panels failed to open and it was placed into a satellite graveyard orbit two years later. TV Sat 2 was successfully launched on August 8, 1989 and was sent to the co-located 19.2 degrees West. But after all the trouble in getting the satellites into position, three out of the four channels were converted from PAL into D2-Mac, the exception being a test card. Worse still, all but one of the channels could be found on either the Kopernikus or Astra satellites in 'old fashioned' but widely accessible PAL. As time went by, the number of channels available on Astra and Kopernikus increased significantly, further denting the DBS schemes.

The debate about D2-Mac evaporated when digital transmission replaced analogue technologies and became a viable and more attractive option. "The system itself did not make a lot of sense, but the pressure was enormous," remembers the former RTL chairman Helmut Thoma. "Today, all this is but a memory."

94/11/01
Astra 1D launched. 18 new channels

94/10/13
In Northern Ireland the two sides agree cease fire

94/11/04
First conference devoted to World Wide Web takes place

94/12/05
Newt Gingrich made US House Speaker

The German Box War

Once upon a time some people imagined the media-rich so-called 'digital super-highway' as a sort of fibre-optic toll road where little toll stations would collect usage fees on the approaches and exits of the magnificent all-available digital highways. Payers would come onto the turnpike; however, ghost drivers and people without a licence to use would not have a chance. When media entrepreneur Leo Kirch started with the conquest of the German television market at the beginning of the 90s, this type of fantasy, of omnipotence dominated the programme creators in the TV industry. Just one security lock – later called the d-Box for short – was supposed to allow access to dozens of digital channels, home banking, tele-shopping, voting and polling as well as providing access to municipal as well as national information.

The decoder box carousel spun around with ever increasing speed.

Those who put the "little box that offers great things" ("kleine Kiste, die Großes bringt") as the catchphrase went, on their TV set were supposed to be able to obtain the ticket to a glittering universe of possibilities. But nothing came of these dreams. The d-Box devoured millions of German marks; the attempt to make access to the market difficult for competitors with closed technical standards turned out to be the wrong track – both financially and technologically. As with the former internet giant AOL, which only allowed access to the net via its own software, holding on to so-called 'proprietary' technology turned out to be a massive flop in the case of the d-Box.

And the story had started out promisingly. The technical requirements for digital television (DVB) started to emerge at the beginning of the 1990s. The driving force behind this was entrepreneur Leo Kirch. The media mogul had already started to

DF1
DAS DIGITALE FERNSEHEN

PREMIERE

1995

95/01/09
Valeri Polyakov completes 366 days in space. A new record

95/01/24
Trial of O J Simpson begins

The set-top box, which was to bring digital TV to the home television, was still in development – but the dispute regarding technical standards and market dominance was already in full swing.

create the framework for digital pay TV in 1991 with his DF1 pay-TV offering. Even as European-backed Digital Video Broadcasting (DVB) technology was in the early stages of its development, the first initiative for the development of digital pay TV took shape. Deutsche Telekom, Kirch and CLT-UFA (Bertelsmann) from Luxembourg founded a company in 1994 for the implementation of pay TV and the related services (MSG). From the programme to the network operation to the decoders, MSG wanted to facilitate access to digital television – and control it with its own technology. The decoders developed for the marketing were to be subsidised and come onto the market in large volumes.

But the key prerequisite was missing. The set-top box, which was to bring digital TV to the home television, was still in development – but the dispute regarding technical standards and market dominance was already in full swing.

A new participant had complicated the TV scene. The offering of free-to-air digital television became available with all public and private channels overnight via the Astra satellites. Astra's considerably cheaper and technically simpler satellite reception allowed decoder prices to fall. It led to beginning of a bizarre controversy, which went down in German media history as the 'Box War'.

Two powerful groups emerged in the struggle for the domination of the television market: Deutsche Telekom, together with Bertelsmann, CLT from Luxembourg and the French pay TV provider Canal+ took on the alliance between ARD, ZDF and RTL with their MSG

A driving force? Formula One was going to be a prime vehicle to promote pay TV sales.

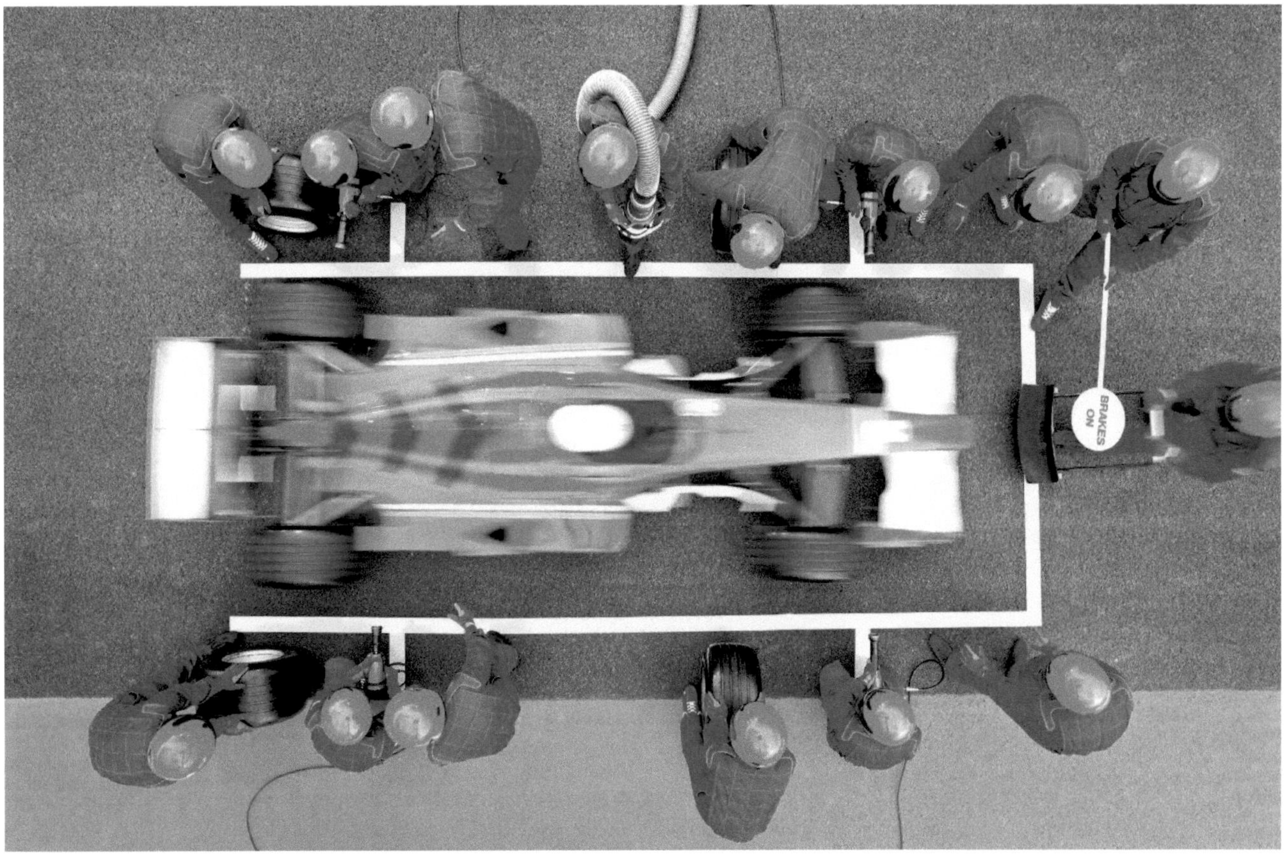

95/02/21
Russia's MIR space station welcomes first Americans

95/02/25
Steve Fossett lands after first solo flight across Pacific in a balloon

platform. No sooner had consumers recovered from the video recording standards battle between BetaMax and VHS before a new TV war broke out. Everyone wanted more or less the same thing – uncomplicated access to the viewer – but using 'their' unique technology.

Independent programme providers as well as public institutions doubted whether the vertically integrated MSG platform would allow them to feed programming into the system under fair conditions and to survive in competition. The decoder manufacturers would have had to adapt their products to the MSG guidelines. Indeed, in November 1994 the EU competition commission confirmed their view and demanded "open and non-discriminating platforms".

With his market-dominating position in digital television, the alliance surrounding Leo Kirch had become too strong for the competition watchdogs. The message from Brussels was that future pay TV competitors would have only had the option of accepting MSG's conditions or staying outside the market. Particularly worrying to the cartel authorities was the planned monitoring of the electronic user guide (EPG). They argued that the programmes of competing pay TV providers could, via the EPG, in this way be placed in less attractive positions.

The Kirch Code

The market forces lined up once again and new forms of cooperation were sounded out. The multimedia operating consortium MMBG was founded in May 1996. Its goal: The development of its own digital platform on the basis of the SECA decoder technology, which was jointly developed by Bertelsmann and Canal+.

Deutsche Telekom was the majority shareholder with 27 percent; the remaining shareholders were the Bertelsmann subsidiary Ufa, Canal+, the CLT Group from Luxembourg, the Daimler-Benz subsidiary Debis as well as the broadcasters ARD, ZDF and RTL. Kirch was also supposed to be brought on board. After months of negotiations he finally stepped up with his own decoder – the d-Box. In September 1996 Deutsche Telekom announced its exit from MMBG; Bertelsmann followed soon after. Collaboration beyond company borders had failed.

Kirch's technology subsidiary BetaTechnik had developed the new d-Box on the basis of its own encryption system. Other systems were only supported after the installation of a so-called multicam. The viewer also needed certain types of smartcards for reception via cable or satellite, which could only be obtained together with a subscription. Kirch wanted to maintain control over the access code for the digital TV network. The first d-Boxes came onto the market at around 1,000 German marks.

Second Attempt

The decoder box carousel spun around with ever increasing speed. In 1997, Bertelsmann, Kirch and Deutsche Telekom made a second attempt at establishing a joint pay TV platform – and the EU once again thwarted their plans. In particular public broadcaster ARD had complained that its own electronic user guide did not function optimally on the d-Box. Because the d-Box has a proprietary access control system, every potential pay TV provider would have to purchase a licence for the use of the access technology, said the decision from Brussels. The Box could even be programmed so that it would be unfavourable for possible competitors. The EU Commission demanded that the marketing of the d-Box be immediately discontinued.

Kirch, after lengthy negotiations, was able to convince Bertelsmann to sell its Premiere share to him. The opposition also regrouped. A loose alliance of free TV providers and equipment manufacturers did not want to give up without a fight and cede the field of digital television to the pay TV providers. Free-to-air television was also to have a chance. At the beginning of 1999 the association 'F.U.N.' (Free Universe Network) emerged from the initiative with the goal of defining an open platform for all programme offerings and bringing it onto the market. Thus, an alternative to the d-Box decoders

Set-top boxes. A huge industry has evolved for the supply of reception equipment

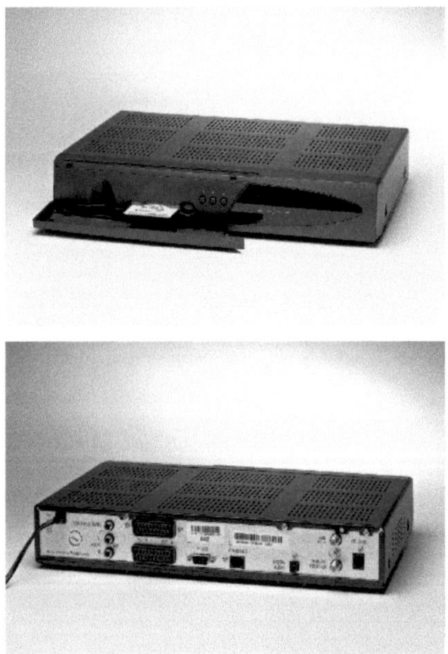

95/03/01
Yahoo is founded in Santa Clara, California

95/03/20
Tokyo subway nerve gas attack. 8 die

Exclusive sports events were a main driver of the pay-TV business from the very beginning

– which were dominated by Kirch both commercially and technically – was supposed to be created for all programme operators.

In 2000 Kirch agreed with ARD, ZDF and RTL on the new MHP (Multimedia Home Platform) standard. MHP was designed to bring television and internet access together. Using MHP Premiere was able to save infrastructure costs on the order of several million German Marks.

And it was the beginning of a single set-top box in which the digital offerings of several broadcasters were packaged. Now electronics companies could begin with the development of MHP-capable equipment. The pivot point was the so-called 'API' (Application Programming Interface) through which the programme providers could distribute their interactive services. It soon became apparent that APIs were key, as were the encryption systems.

The interest in pay TV – which is not excessive in Germany – was served by Premiere via the d-Box while digital free TV via satellite increasingly developed into an open market with decoders from independent manufacturers. Only after the insolvency of the Kirch Group and the technical reorientation of Premiere, today Sky Deutschland, was the system competition against the Box relinquished with a new Premiere decoder.

95/04/19
Oklahoma City bombing:
168 people killed

95/05/01
Jacques Chirac is elected
President of France

07

Astra Cracks the German Market

07
Astra Cracks the German Market

Jörn Krieger

Only a few households in Germany witnessed the event which changed the nation's broadcasting landscape forever: On December 8, 1989 three analogue transponders on Astra 1A (19.2 degrees East), which had been carrying Astra test captions for months, suddenly sprang to life. Germany's main commercial television channels RTL plus (now RTL), SAT.1 and PRO 7 (now ProSieben) appeared, broadcasting their regular programmes. And they could simply be picked up with a conventional analogue DTH satellite system on a common TV set in PAL picture standard. No special LNB, converter or decoder box was necessary. This was quite unusual in Germany's young DTH market and one of the reasons for Astra's fast rise in popularity. The move laid the cornerstone not only for the development of Astra into Germany's leading DTH satellite system, but also for conventional analogue television in PAL as the broadcast standard for DTH transmissions.

The appearance of German channels on Astra was a breakthrough which marked the birth of commercial television in Germany. Helmut Thoma, founder and long-standing chairman of RTL, had started to produce programs in a converted bus garage in Luxembourg – "very improvised and with very modest technical equipment," as he recalls. "We urgently needed distribution and spoke to Astra, but on the other hand, did not want to alienate anyone" he says. "There was huge pressure from Germany not to go with the Luxembourg initiative which was backed by the Americans. And then, thank God, SAT.1 took the decision and signed with Astra, and thus the dam was broken. Leo Kirch had gone it alone; we thought that they would be more considerate of the German Federal Post but they actually did not care, and therefore, we were also able to make such a quick decision to go with Astra."

95/07
Amazon starts trading

95/08/24
Microsoft releases Windows 95

C. Forrester (ed.), *High Above: The Untold story of Astra, Europe's leading satellite company*,
DOI 10.1007/978-3-642-12009-1_7, © Springer-Verlag Berlin Heidelberg 2011

The appearance of German channels on Astra was a breakthrough which marked the birth of commercial television in Germany.

Sophia Loren brightens up Thomas Gottschalk's 'Haus-Party' on Sat.1

Jürgen Doetz, at the time head of German broadcaster SAT.1, says that he was involved in talks with Astra's Pierre Meyrat very early on in the process. "And for the SAT.1 channel I was then able to be the first German operator to sign a contract on the use of a transponder. Only hours before our planned press conference, RTL plus, as the Luxembourgers still called themselves at the time, also signed – as the second channel."

Before being joined by the German broadcasters, Astra 1A transmitted channels from countries mainly interesting to expatriates from those regions, for example from the UK, the Netherlands and Scandinavia. However, with Eurosport, MTV Europe, Screensport and Lifestyle, Astra 1A was already the home of several pan-European broadcasters. Germany's domestic television channels were spread across a number of satellites on different orbital positions including TV-SAT 2, DFS Kopernikus, Eutelsat and Intelsat.

Viewers wanting to access the whole range were left with no choice but to invest into a motorised dish and specialised equipment as transmissions took place in various

95/09/03
eBay founded in San Jose, California

95/09/10
US Open, Sampras beats Agassi

RTL's 'Tutti Frutti' TV show, hosted by Hugo Egon Balder, was probably responsible for selling more Astra dishes and receivers around Europe than any high-profile movie or sports content! Tutti Frutti was the German version of an Italian game show (Colpo Grosso) and ran for 3 years from 1990 and some 140 episodes. British audiences, restricted by very conservative broadcasting policies, couldn't get enough of Tutti's pretty ('Cin Cin') girls, their partial nudity and 'stripping housewives' segment. Needless to say it caused outrage amongst Britain's chattering classes, which only further promoted dish sales! Incidentally, Tutti Frutti was an early pioneer of 3D transmission using the clever 'Pulfrich effect' and where extra depth was given to viewers on an ordinary 2D screen

The political aim was to maintain control over the countries' broadcast airwaves, based on the historically tight regulation of terrestrial infrastructure and cable networks.

95/10/19
Astra 1E launched. Another 18 channels

95/10/03
O.J. Simpson found not guilty of double murder

standards. Germany's domestic direct broadcast satellite (DBS) TV-SAT 2 used the D2-MAC system, a mixed analogue/digital technology backed by Germany and France, broadcasting on BSS frequencies for which a special LNB was needed. DFS Kopernikus transmitted in the conventional analogue system, but used a combination of 11 and 12 GHz frequencies, again requiring a special LNB for receiving the whole line-up. The satellites operated by Eutelsat and Intelsat used the 11 GHz band, but didn't carry many German channels. Also, because the satellites had different transmission powers, dish sizes required for reception ranged from 45 cm for TV-SAT 2 to 180 cm for Intelsat.

In other words, it was a right mess. Only a handful of enthusiasts, so called DXers, had the equipment, knowledge and interest to get involved in DTH reception. The move of SAT.1 and RTL to Astra changed that. Further channels such as 3sat, VOX, RTL 2 and Super RTL followed and, helped by falling prices for reception units, 19.2 degrees East soon became Germany's fast-growing DTH 'hot spot'. While in November 1989 an estimated 55,000 households received the DSF Kopernikus satellite and only 25,000 had their dishes pointed at Astra, the Luxembourg based satellite system managed to catch up with its Telekom competitor only two months later. A total of 850,000 satellite reception systems were sold in Germany in 1990. At the end of 1990, more than 70% of satellite dishes in Germany were positioned to Astra 1A which back then carried 16 channels. Telekom's DSF Kopernikus had nine and TV-SAT 2 only 4 channels onboard. The variety of channels, inexpensive reception units and easy installation were the keys to Astra's growth.

The unexpected success of the Luxembourg-based satellite system gave state-controlled telecommunications company Deutsche Telekom cause to worry: Could it be that Germany's domestic satellite ventures TV-SAT 2 and DFS Kopernikus would soon become insignificant? At the time, Europe's largest countries including Germany, France and the UK tried to establish domestic satellite systems. The satellites transmitted with high power in the BSS band, so reception was possible with a dish size of only 45cm, compared with the 60cm required for Astra. But they didn't have room for many channels. Heavy regulations stipulated which channels were allowed on the satellites and which technical parameters could be used. The political aim was to maintain control over the countries' broadcast airwaves, based on the historically tight regulation of terrestrial infrastructure and cable networks.

But one point was not given enough thought: Satellite signals don't stop at national borders. While the BSS satellites were restricted to domestic channels, Astra became the

In space, the idea of a borderless Europe was already well alive long before boundaries were opened on the ground.

95/11/22
'Toy Story' released by Pixar and Disney

96/12/20
Carl Sagan, astronomer, dies

95/12/25
American actor and singer Dean Martin dies

home of a mixture of channels from different countries which soon became popular among viewers. In space, the idea of a borderless Europe was already well alive long before boundaries were opened on the ground. Also, it was much easier for a broadcaster to secure space on Astra compared with the domestic satellites which had always been subject to political interests.

The many foreign-language channels on Astra from other European countries soon became popular among German students and other people interested in brushing up their language skills. As an additional bonus, UK based channels such as Sky One showed current episodes of popular US series such as 'The Simpsons' long before they appeared on German television, giving satellite viewers an exclusive peek into what 'ordinary' viewers would have to wait months to see.

Another boost for Astra's popularity in Germany came from the country's reunification. As soon as the borders were opened in November 1989, East Germans, hungry for both unbiased political information and entertainment, stormed retail shops in Germany to buy DTH satellite reception units. In 1990, a total of 350,000 DTH systems were sold in eastern Germany with almost all purchasers opting for the inexpensive, easy-to-install Astra kits. This provided German commercial broadcasters with a most welcome ratings boost, but left the public broadcasters in doubt: Was it still the right decision to rely only on the country's domestic broadcast infrastructure? ARD and ZDF finally joined Astra with their main channels in 1993. At last, 19.2 degrees East had all of Germany's main channels on board – and the fate of Germany's national DTH ambitions was sealed. In 1994, Deutsche Telekom became a 25% shareholder in Astra.

With more capacity becoming available due to the addition of further satellites to the 19.2 degrees East fleet, ARD's regional channels joined the platform, first from large federal states, followed by smaller states. This enabled viewers living in the South to see what was going on in the North and vice versa and made it possible to watch familiar programmes even while on interstate holidays, further contributing to Astra's popularity. With the growing number of DTH kits sold, prices dropped, making satellite reception affordable to most households. Satellite equipment wasn't only sold in electronics shops any more, but widely available in supermarkets, hardware stores and large retail chains. Self-installation of the set

Berlin, November 1989: Stepping from West into East

1996

96/01/29
Fire destroys La Fenice, Venice's opera house

96/02/10
Chess computer 'Deep Blue' beats Garry Kasparov

without the need for a technician was another key to quick market penetration.

With the country's main public and commercial channels on board, Astra soon developed into a serious competitor for the domestic cable network which back then was in the hands of Deutsche Telekom. From mid-1980, Telekom poured millions of Deutschmarks into the deployment of a nationwide cable network, financed by taxpayers' money – and now saw the investment in jeopardy. Instead of cable, a growing number of households aiming for multi-channel television opted for DTH satellite reception. Also, existing cable households cancelled their contracts and installed a satellite dish.

"In 1991, individual dish fees were abolished, thus liberating broadcasting."

The offer from the sky was tempting: DTH was free once the system had been installed, while cable television incurred a monthly fee. Satellite also has a larger number of channels and much more capacity. While Telekom often had to transmit two channels on a single slot in time-sharing mode on its cable network due to a lack of capacity, satellite viewers enjoyed both channels' full 24-hour schedules. This was often confusing to cable households as TV programmes were announced which they couldn't see because another channel used the slot at the same time. Also, in some cases switching took place at different times during weekdays and weekends – with no one really knowing what would happen on nationwide or regional public holidays.

In a bid to keep Astra's prospects in check, Telekom declared Astra 1A to be a telecommunications satellite for which, in contrast to DBS satellites such as TV-SAT 2, a reception fee applied. This was despite Astra 1A only carrying television channels and no telecommunications traffic such as telephony or data transmissions. Households wishing to receive the signals had to register at Telekom and make a one-off payment of 50 Deutschmarks in addition to a monthly fee of 5 Marks. After September 1, 1989 the monthly fee was waived and the one-off payment reduced to 25 Marks, but registration was still required. The controversial registration and fee were dropped on February 1, 1991. "At that moment," remembers the former Post Minister Christian Schwarz-Schilling, responsible for the cable expansion and also the liberalisation of the broadcasting landscape, "the normative power of the factual had a big push through Astra."

Astra's alarming success also provoked Telekom to launch a movies and series channel in conjunction with commercial television broadcaster ProSieben, Der Kabelkanal ('The Cable Channel'), which was exclusively available on its cable network in 1992. Even the satellite feed on Eutelsat to serve cable networks was encrypted, shutting out any potential satellite viewers. It only took until 1995 for the channel to launch on Astra. Management couldn't afford to ignore the massively growing DTH audience any longer. Now renamed kabel eins, the word 'Kabel' is the only remainder of the channel's past as a satellite spoiler.

Not only Telekom, but also competing satellite operator Eutelsat tried to put a halt to Astra's growth. However, attempts by the Paris-based company to promote its 13 degrees East position in the mid-90s as a DTH 'hot spot' in Germany were met with limited success. Even large marketing campaigns with music channels Viva and Viva Zwei, which back then were available exclusively on 13 degrees East, didn't help. In 2000, both channels joined Astra.

96/03/13
Dunblane school massacre.
16 students and 1 teacher killed

96/03/20
First UK cases of Mad Cow
Disease revealed

08

Overcoming the Barrier

08
Overcoming the Barrier

Helmut Thoma, Chairman of RTL Television 1984-1998

Helmut Thoma, former Chairman of RTL Television, says that the situation that existed in Germany as regards satellite transmission was controversial from the very start. "The Social Democrats who governed in coalition with the Liberal Party FDP until 1983 were strong proponents of public broadcasting, whereas the Conservatives on the opposition regarded themselves as subject to discrimination by the public broadcasters. There was the famous term 'Schweigespirale' (a spiral of silence) that was coined by Elisabeth Noelle-Neumann. The political battle lines were fairly clear. In France, ironically, the situation was exactly the reverse: The Conservatives were in power, and the Socialists under Francois Mitterrand were in favour of opening the system to private broadcasters and even helping to create some broadcasting stations."

96/04/09
Astra 1F launched from Kazakhstan.
First-ever 'Western' launch

96/04/03
Unabomber arrested in US

96/04/14
Nick Faldo wins 60th Golf Masters

C. Forrester (ed.), *High Above: The Untold story of Astra, Europe's leading satellite company*,
DOI 10.1007/978-3-642-12009-1_8, © Springer-Verlag Berlin Heidelberg 2011

Thoma insists that even if two satellite operations had been in place there was no guarantee that either would have succeeded. "Satellite TV was a trigger for private TV and enabled it to expand more rapidly. That also applied, for example, after the fall of the Berlin Wall, when Western television could then suddenly be received in even the remotest corners of the GDR."

"Cable was not an issue for us at RTL – even though we did not even have the Astra satellite then, but only the direct broadcasting satellite that could be received with relatively small dishes. Relatively small meant a diameter of less than one metre at the time. Luxembourg, in other words RTL, would have received a channel on this direct broadcasting satellite. But then the satellite did not open one of its solar panels after the start, so that it ended up as space junk, making all plans obsolete. Cable only arrived after that – along with Astra, which could be used to feed the cable stations. The advantage of Astra was that it was classified as a communication satellite, as if one were to install a directional radio route in space. It was not regarded as a broadcasting satellite."

"After the Wall came down (...) people used the money they received on Monday to go out and buy first a car and then a satellite dish on Tuesday."

Asked whether the Luxembourg scheme wasn't just an attempt to get around the rules, Thoma said: "German Federal Post had difficulties coming to terms with the Astra satellite and did all it could to prevent Astra. They even wanted to launch their own satellite system, which never came to anything, and then they joined Astra as shareholders themselves later on. German Federal Post was still a government ministry at the time and did not exactly win laurels in this matter. Nonetheless, all that didn't help, so that things ultimately evolved the way they did. When Murdoch then also talked to Astra to supply Britain, it proved that there was simply no way to get around Astra. That was the real breakthrough. There was simply no successful method for preventing that from happening. It was simply like voting for a new technology with your feet."

Thoma played an important role in the process, and one that also helped RTL at the same time. "I was there immediately after the Wall came down and looked at the whole situation. It really was the case that people used the money they received on Monday to go out and buy first a car and then a satellite dish on Tuesday. I saw that in what was then called the 'Valley of the Ignorant' in Dresden because it was outside the western broadcasting range. The big apartment blocks there were all covered with satellite dishes. It looked as if they had all been attacked by huge mussels. Cable could never have done that in such a short period of time!"

"If you consider that all European countries are now covered by satellite television, where in the past people only had their national television, you can see what a powerful role TV has acquired and how omnipresent it has become."

"The satellite also offers tremendous advantages. For example, if we look at energy technology. The start uses a lot of energy and causes major carbon emissions. But once the thing flies, it uses sun energy, in other words renewable energy, and generates very low emissions. This is practically the best method for transmitting programmes across large regions."

96/05/31
Benjamin Netanyahu elected Prime Minister of Israel

96/06/04
Ariane 5 explodes 40 seconds after takeoff in French Guiana

Jürgen Doetz Managing Director, SAT.1, Germany 1985-2004

Jürgen Doetz explains that satellite was already playing a role in Germany at the launch of private TV in 1984. "Satellite had first become possible under constitutional law through cable technology; the first private German broadcaster, SAT.1 launched on January 1, which explains this, with a so-called ECS-Licence (European Communication Satellite) – a licence for the use of a transponder on a communications satellite for the "connection of the cable islands in the Federal Republic of Germany".

"But it also became clear very quickly at that time for us that the growth of cable reach alone would take a very long time for economically successful private television. We had to get out of the 'cable ghetto', but a communications satellite was completely inappropriate for direct supply to the households because of the necessary size of the parabolic dish – the planning and hoping was concentrated on direct satellites; no one had seriously thought about terrestrial frequencies at the time."

"The technology first created the basis for the pioneers of private television. Cable was initially in the fore in Germany, but satellites were also part of the strategic plans of the major players from the beginning. Bold projects were in discussion, for example, I remember very concrete conversations with Leo Kirch and head of ORF Bacher and the 'Bertelsmann players' from Gütersloh. So that just means: The satellite was always present and was always crucial for national operators to be able to actually implement a 'dual broadcasting system' in Germany. With cable, the official project strategists always envisioned 'islands' and a technical reach of five million households as a sufficient basis for the financing of a full programme channel – which was, of course, never taken seriously by most entrepreneurs. And for that reason, after the failure of the German direct satellite, the provision of the first Astra satellite from Luxembourg became the deciding signal that private television providers really had a chance to be able to create a dual broadcasting landscape in Germany with cable and satellite."

Asked whether the German authorities reacted to satellite technology, and Jürgen Doetz stresses that there were differences: "Of course, many people initially think of the deployment of Hessian policemen when satellite reception was supposed to be demonstrated in a hotel in Frankfurt – this is also one of my unforgettable memories. But it remained an episode – the debate about the disfigurement of blocks of flats as a result of the dishes on balconies, but that was all dealt with relatively quickly."

"In my opinion it is more important to acknowledge – even today – the project to place a German direct satellite in orbit. The channels on the first satellite were distributed in terms of media policy – you can see the repercussions of these decisions even today – and then the disaster came about. We went to sleep happy after the celebration of the successful rocket launch, only to wake up the next morning with a satellite in orbit, whose solar panels, which are vital for the energy supply, had not opened. And after even Minister Schwarz-Schilling, who always fought and argued for communications technology, was not able to beam himself into space to repair the paddles, the whole 'German direct satellite' project was seen to have failed. And then the clever Luxembourgers moved into this market niche and the German government did well to not put up resistance with a new project for some

96/07/19
1996 Summer Olympics in Atlanta begin

prestige purpose. The last 25 years proved it right!"

"From day one of private television in Germany, which – and there is no denying it – in terms of entertainment law and media policy was only made possible by the entry of cable technology in Germany, the satellite was supposed to and had to – according to our expectations and strategies – create the critical prerequisite for a national spread of SAT.1 with the viewers and thus to become economically successful. It is not difficult to recognise the significance in the name of our channel, which we had decided on when almost no one in Germany had direct-to-home reception on the agenda."

Doetz says that he was involved in talks with Astra's Pierre Meyrat very early on in the process. "And for the SAT.1 channel I was then able to be the first German operator to sign a contract on the use of a transponder. Only hours before our planned press conference, RTLplus, as the Luxembourgers still called themselves at the time, also signed – as the second channel."

"When we put on an advertising tour throughout 34 cities of the former GDR with SAT.1 in the year of the reunification, the town squares were filled with 'Wheel of Fortune' fans – many people had made their own satellite reception equipment in order to be able to watch the private channels."

He describes that satellite's pioneering spirit was not unimportant. "First of all you have to remember that in spite of Europe-wide distribution of the satellite channels, the national market was always crucial for the individual broadcasters – and continues to be today. However, for these broadcasters 'Europe' also had consequences for regulation in view of the transnational possibilities for reception – the corresponding 'television directives', for example for commercials, have their reasoning here.

For the strategy of European-oriented operators – and RTL was this right from the start with its roots in Luxembourg – satellite television was a precondition for a successful corporate policy."

Satellite was also key for German re-unification. "Satellite reception was crucial for the high acceptance that our broadcasters still have today in eastern Germany – especially for us 'private broadcasters' – after, with the exception of large parts of Saxony, the population was at least technically provided for terrestrially with ARD and ZDF in the time of the GDR. But that was too little for many people: When we put on an advertising tour throughout 34 cities of the former GDR with SAT.1 in the year of the reunification, the town squares were filled with 'Glücksrad' (Wheel of Fortune) fans – many people had made their own satellite reception equipment in order to be able to watch the private channels. I of course do not want to allot us a similar role to that of ARD and ZDF with their images for example from Leipzig, which were transported back into the GDR via a terrestrial path, but after the reunification, satellite reception was the best basis for the provision of all households with all German channels," says Jürgen Doetz.

"Satellite television first allowed a dual broadcasting order to become reality in Germany, and today it is the crucial motor of a development, which is making a new media order absolutely necessary for Germany! Astra did its homework – now it is up to the equipment industry, the regulators, the 'cable people' and us programme operators, to overcome the obstacles in the market. Here HDTV plays an important role as 'draft horse', which is significantly documenting the digital added value – and thus of course for the design of the new media landscape, of the digital world."

Dieter Stolte Director Zweites Deutsches Fernsehen (ZDF), Germany 1982-2002

Dieter Stolte says technology has always been an important driver for broadcasters. "That applies especially to television. Multi-channel sound, colour TV, cable channels, satellite transmission, HDTV, JPTV – these are all technological developments, invented and driven by brilliant engineers. Political regulation and programme development always arrived afterwards to give the technology a legal framework and use the new opportunities it provides. The judgment of the Federal Constitutional Court in February 1961 made the introduction of private television in Germany conditional on adequate supply of the public broadcasters with terrestrial frequencies. Their aim was full supply of all German TV households. Under the ZDF state treaty of June 1961, ZDF frequencies were allocated by the German Federal Post, while the ARD largely procured frequencies itself and distributed them regionally. Private television, therefore, only had minimal chances of frequency access."

But cable and satellite television soon changed that landscape. "Federal Post Minister Christian Schwarz-Schilling made the start when he decided to fit metropolitan areas with broadband cable and operate the ECS satellite," says Stolte. "The latter had Eastern and Western beams. One went to the private broadcaster SAT.1 and the other to the public-law ZDF to develop a television

1997

97/01/09
Solo yachtsman Tony Bullimore
found alive after 5 days

programme together with ORF and SRG for the German language and culture region. 1984 saw the birth of 3sat, which has now celebrated its 25th anniversary. To sum it up: The early Eighties were an ideal period for the introduction of satellite TV, and that was also urgently needed."

In Stolte's view, Germany's dual media system definitely needed satellite television to gain an equal playing field for dissemination of its programmes. "In addition, only this new technology could bring into being and guarantee the free flow of information required by the European community of nations."

"Satellite technology corresponded ideally to the political spirit of the age and the development of democracy in a free global community."

Dieter Stolte says that German telecommunications regulators could have acted a little speedier. "The slow speed was due both to the federal broadcasting structure in Germany and to the ideological controversies in German media policy."

Asked what he expected from satellite technology at the time, Stolte says the ZDF always had a strategy that was technically informed and anticipated developments early on. "We were always looking for opportunities to compensate our competitive disadvantage compared to the combined force of the nine state broadcasting stations of the ARD with their multitude of radio and TV programmes by being faster. The creation of Astra and its dynamic development therefore matched our own interests. We were aiming for the best. We made use of our opportunities. It was like a game of cards: you can always only play the cards you get, but then need to do that wisely and with a clear goal in mind. The ZDF was the pacemaker at 3sat, Phoenix and KiKa and ultimately also Arte, the four most important creations of public-law TV during the Eighties and Nineties."

"Until satellite TV was introduced, European media was very much defined by regions and language territories. This even extended to the procurement of transmission rights, which was exclusively territory-based. That applied especially for sport broadcasting rights and blockbusters from Hollywood. In addition, international politics both in the European Community and worldwide was committed to the free flow of information under the Helsinki Accords (1975). In short: satellite technology corresponded ideally to the political spirit of the age and the development of democracy in a free global community."

Stolte argues that the reunification would have been inconceivable without television in general and satellite television in particular. "The continuous presence of a freedom-oriented world view compelled citizens in Eastern Germany to make comparisons and showed that the propaganda system of the GDR produced only permanent brainwashing and lies. Satellite technology acquired prime importance directly following reunification. In only a few months, the former citizens of the GDR reequipped their TV sets. At a stroke, RTL, SAT.1, PRO7, VOX and other commercial channels could be viewed everywhere. That triggered a wave of growth. The private broadcasters with their perky and frivolous entertainment formats quickly overtook the public-law broadcasting programmes. To this day, their market shares in Eastern Germany remain higher than those of ARD and ZDF."

"Satellite broadcasting has changed the media landscape fundamentally," says Stolte. "First of all, no programme provider can operate today without satellite transmission; secondly, all programmes are viewed across national borders, which is especially pleasing for German holidaymakers and residents in the Mediterranean region; thirdly, all competitors have equal opportunity as regards technical reception capability."

97/02/09
'The Simpsons' 167th episode.
Longest-running cartoon series

97/03/27
'Heaven's Gate' cult mass suicide
in California

09

Astra Adds Up for Canal+

09
Astra Adds Up for Canal+

Serge Siritzky

To understand the evolution of television in France, one must understand that the state has always wanted to control it. For a long time it was mainly to control news on the most powerful media. Then – and it is still the case – it was to favour French production and French technology and, if not directly favour, at least influence pan-European production and technology.

Up until the 1980s, there was only one way to broadcast television: through a terrestrial network and with a limited number of frequencies, thus a matching limited number of channels. In line with its industrial policy, the state had chosen to broadcast television by using a French technology, Séquentiel Couleur à Mémoire (Secam), while the rest of Europe had selected the Phase Alternating Line system (PAL). There were just three national state-owned channels, financed mainly by a licence fee and also by advertising. But the use of advertising was limited as a show of goodwill – and protection – for the press. Those public channels were tightly controlled by the government.

Suddenly, in 1982, TDF, a public company that had the monopoly over the transmission of TV signals in France, decided it should do something with the VHF network freed up by broadcaster TF1 which had migrated onto UHF frequencies. TDF revealed that there could be sufficient frequencies available for two more national channels, covering at least 60% of the population.

At the same time other ways to broadcast television were appearing internationally. First, cable which could transport a great number of television channels. Cable had expanded in the United States where, because of skyscrapers in urban areas, but also in order to get signals to far-away suburbs, co-axial cable was used to deliver clearer images. Charles Dolan was the pioneer, in 1965 building a cable system in Lower Manhattan, in New York. In the United

97/03/30
Channel Five begins broadcasting in UK

97/04/05
Allen Ginsberg, beat poet, dies at 80

C. Forrester (ed.), *High Above: The Untold story of Astra, Europe's leading satellite company*, DOI 10.1007/978-3-642-12009-1_9,

States, cable initially carried the three main networks and a handful of local stations which varied depending on location.

HBO, the USA's first 'pay cable' channel, launched in 1972, showing movies and speciality sports (usually high-energy boxing matches). Even this system wasn't nationwide in that playout tapes were sent to individual cable head-ends. HBO went onto satellite on Sept 30, 1975, when it transmitted the Muhammad Ali vs Joe Frazier 'Thriller in Manila' boxing bout. Ted Turner, the owner of a local television in Atlanta, TNT, turned his local station into a national broadcaster by sending his signal to the head-end of the major cable systems through the use of a telecommunication satellite. The cable head-stations were equipped with dishes of more than 1.2 metres in diameter.

"The French government understood that it wasn't possible to keep limiting the number of channels to three."

The French government understood that it wasn't possible to keep limiting the number of channels to three. But as Françoise Giroud, the well-known Managing Editor of the weekly 'L'Express' had written: "When it comes to television nothing is impossible to French politicians, except good decisions."

The government didn't want to create more public stations, financed by more taxes. It thus decided that the fourth terrestrial network would be a movie pay television, financed by subscription, like HBO. Thus, Canal+ was launched in 1984. It was a subsidiary of Havas, the large French advertising agency, and also controlled by the state. Two years later, two private national televisions, financed by advertising, were authorised to be broadcast. And at the same time, the government decided to let the major urban areas build cable systems and also allow the launch of TV satellites. But all those decisions were taken separately, without any sophisticated overview.

In France, households could receive five national channels (three public and two private) and one pay television channel (Canal+). Cable had nothing to carry and besides the construction of its network, the investors had also to finance any new channels. Its monthly subscription rate was

97/05/02
Tony Blair wins UK election

97/05/30
JK Rowling's 'Harry Potter and the Philosopher's Stone' published

France's César Awards, a huge TV ratings favourite

thus very expensive, 2.5 times higher than in Germany, for example. With this sort of fee, the maximum penetration rate was estimated to be just 20 to 25% of homes, which was far below the 50% needed to refinance the cable network. It was, coincidentally, also the penetration rate that Canal+ finally reached. Pay TV would eventually turn out to be financially profitable. French cable could manage a technical reach of just 50% of households, and thus the simple formula of 20% of penetration of the 50% of homes passed equalled just 10% of homes and was simply insufficient for individual cable channels to achieve financial breakeven.

At the same time, the French government decided to launch TV satellites. Telephony penetration in France was very low and the administration in charge of this sector wanted to concentrate its investments on boosting the telephony take up. The French PTT decided it shouldn't spend time or money on television. However, TDF, the publicly-owned company which handled terrestrial transmission, was chosen to set up a project for a television satellite. Being a monopoly, TDF had no marketing policy and did not try to see who its clients might be. Besides, at that time, it was far from clear – at least in France – whether telecommunication satellites could reach households directly.

Some industrialists said the TV signal broadcast by telecommunication satellites was not powerful enough to be caught by a small dish (smaller than 80 cm). TDF agreed with this and therefore ordered satellites with powerful transmission capabilities (of 240 W). The Germans made a similar choice, as did the British with their pair of Marco Polo satellites. But nobody had experimented with these satellites. As they were powerful, they were difficult to cool, and thus operationally fragile and needed on-board back-up transmission equipment.

But there was another challenge: Besides using powerful transmitters, these satellites used a new broadcasting standard: D2-Mac (see 'Europe's Dish and Box Wars', Chapter 6). Companies like Philips and Thomson in Europe, and Sony in Japan, were convinced that the future of television was in higher definition and the 16:9 widescreen format instead of the so-called 'square screen' which followed the old movie format of 4:3. The Japanese government had decided to back Sony's MUSE analogue high-definition standard which used a 16:9 format screen. The European decision to support the D2-Mac concept was – at least partly -

> The French and German satellites had one technical failure after another and together could only broadcast four signals, and soon it became apparent that they had little future.

to avoid paying royalties to Japan, as well as supporting their own manufacturing sector.

The problem with MUSE and D2-Mac television sets (and set-top boxes) was that they were more expensive than PAL or Secam versions. Production volumes were small, and that made the unit price quite significant. There was an obvious snag in that the transmitted images were in 16:9 while most of the TV programmes were still in 4:3 (as were the TV sets). And the quality of the image was not dramatically better than that of PAL or Secam.

At the exact same time as the French and German governments were stubbornly backing the high powered D2-Mac satellites, Astra started to launch satellites. These satellites used a technology drawn from conventional telecommunication satellites and using uncomplicated PAL technology. Each satellite could carry 16 transponders that could be received by small dishes (no larger than 60 cm). And it didn't need spare transponders or spare satellites. TDF and its German counterpart said it was impossible to receive the Astra signals with such small dishes, and besides it was altogether too risky not to have spare transponders!

But perhaps the greatest difference between Luxembourg's SES and France's TDF was that the TDF management had little or no marketing culture. The French satellites were called TDF1 and TDF2 and the Germans Sat1 and Sat2. They had the same location in the sky. But TDF1 was launched in 1989, before having signed any contracts with clients and before really knowing what channels to put onto the satellite. The only logical client was Canal+, which wisely wanted to reach households in areas which were not covered by its terrestrial broadcasts. Canal+ signed a contract with TDF and rapidly achieved 50,000 subscribers through TDF1. Unfortunately, the French and German satellites had one technical failure after another and together could only broadcast four signals, and soon it became apparent that they had little future.

In France, besides Canal+, there was demand from some broadcasters that had previously been depending on cable with its low penetration. Although those channels had many different owners, the leader of the pack was Canal+, which wanted to control the marketing of a bouquet of low-priced pay TV channels over satellite.

Canal+ had to choose which satellite operator to use. On the one hand there were the Astra satellites, on the other those of France Télécom, the public company that replaced the DGT, the telephone administration. France Télécom had three satellites, Télécom A, B and C. Each one could carry 11 channels. The problem was that they had not been conceived for television and were not located in the same place: one needed two dishes or a dish with two LNB pick-ups to receive the different signals.

The President of Canal+, André Rousselet, not unreasonably had a preference for a French solution, and using France Télécom's satellites. But he had first to change the government's policy towards D2-Mac, which he believed was at a dead end. In September 1991, the Federal Communications Commission (FCC), the US broadcasting authority, had decided that the future of television was in digital broadcasting. This would enable satellite operators to multiply by 4 to 8 the number of signals broadcast in the bandwidth occupied by a single channel. And Canal+'s shrewd management was convinced that the FCC was right. After two years of discussions, the French

97/08/14
Timothy McVeigh sentenced for Oklahoma bombing

97/08/31
Diana, Princess of Wales, dies in Paris

government, in 1992, was finally convinced that D2-Mac was dead, much like Japan's MUSE analogue high-definition system. In 1992, Canal+ decided to choose a Télécom satellite to broadcast its Canalsatellite bouquet, mostly because France Télécom was a French company and although France Télécom was not much more market-oriented than TDF (it too had launched its satellite dedicated to television before having signed any contracts). The subscription price of Canalsatellite was much the same as Canal+ or cable fees, and thus relatively high.

France Télécom promised it would launch a satellite that would carry digital signals by 1997. But Astra promised a digital satellite as soon as 1995 and added that by 1997 it would have five satellites, each with 16 channels, out of which two of the satellites would address the new, and very efficient, digital transmission system. The Canal+ group, already profitable, had expressed plans to expand into Spain and Italy. For those markets Astra seemed, by far, the most competitive satellite operator and, in November 1994, Canal+ contracted four transponders on the planned digital satellite, Astra 1D. And starting to work closely with Astra, Canal+ management realised that Astra was much more dynamic than France Télécom, and also that it should not wait until 1997 to go digital in France. And France Télécom was somewhat blind to the fact that it had a powerful competitor on the French market. In 1995 Canal+ rented more transponders for Canalsatellite on Astra 1F, which would be launched in 1997.

The strategy of Canal+ was initially to follow the pattern of HBO in the United States: to offer to its growing number of subscribers a service based not on one movie and sport premium channel, but several channels. Besides this premium channel a subscriber

Claude Makélélé Sinda is captain at Paris Saint-Germain, and a world-class defensive midfielder. He has also played for Chelsea and Real Madrid. Some 11% of France's population supports PSG, and helps ensure their games are a major TV event on Canal+.

97/09/05
Mother Theresa of Calcutta dies

97/10/27
Stock markets crash

97/11/11
Mary McAleese elected the 8th President of Ireland

would, for the same price, receive a movie channel, a sport channel and a family channel. For instance, the movie channel would supply more reruns of the 'premium' channel, but also more specialised films that could appeal to different categories of movie buffs. The acquisition price of those additional films was low, but the offer widened the number of potential clients. The same argument applied to speciality thematic channels. Thus, for a limited increase of is programme and transmission budget, Canal+ managed to strongly widen its potential subscriber base. Besides, more transponders would enable Canalsatellite to increase the number of channels it offered while reducing the cost of their transmission.

France Télécom promised it would launch a satellite that would carry digital signals by 1997. But Astra promised a digital satellite as soon as 1995 and added that by 1997 it would have five satellites, each with 16 channels.

At the end of 1994, Canal+ and Canalsat had 200,000 subscribers through the Telecom satellite. Two years later they had a total of 400,000 subscribers, out of which about half were through Astra's satellites using digital transmission. At the same time, French cable had a total of two million subscribers and Canal+ itself had some three million subscribers, mostly through the initial analogue terrestrial network.

At about this same time the free terrestrial networks, including privately-owned TF1 and M6, and the publicly-funded networks France 2 and France 3, understood that satellite would become more and more important, even for channels already reaching viewers via terrestrial antennas: Households would inevitably have a satellite dish that enables them to get more and more channels and viewers would end up mostly watching signals they received on the satellite dish. Thus, as the number of viewers to Astra's specialised channels increased, the ratings of 'terrestrial' networks would inevitably decline if they were not on Astra. And, in any case, the general opinion was that if there were to be more and more specialised channels, then mainstream broadcasters should also be present in the growing specialised channels market.

Discussions started between Canal+, TF1, M6, France 2 and France 3 about broadcasting on satellite. Canalsat believed it could impose quite tough terms; the Canal+ management seemed not to pay that much attention to the fact that Europe's telecom monopolies had created a new European satellite company in the shape of Eutelsat, and Eutelsat was building new satellites and wanted a portion of the television market.

Consequently in 1997, TF1, M6, France 2, France 3 and France Télécom decided to create a company that would launch a television bouquet on a Eutelsat craft. The consortium would have to create many new specialised channels as well as movies and sports premium services in order to offer an alternative to Canalsat and Canal+, as Canalsat had shrewdly made exclusive agreements with most of its existing specialised channels.

The concept was a very risky bet. The new bouquet was called Télévision Par Satellite (TPS). Its management believed it would have two advantages: first to broadcast the existing network channels of France's public broadcasters TF1, M6, France 2 and France 3; second, it would have a lower subscription price than Canal+, Canalsat or cable.

At the same time, an important French independent producer, Claude Berda´s AB, which owned a large library of 12,000 hours of French, European and American programmes, decided to launch its own satellite bouquet on Eutelsat. It would be a 'low-cost' bouquet, as most of the programmes would be reruns coming from its library. And it would carry the only French channel devoted to 'adult' material. Its subscription rate would be much lower than TPS.

97/11/12
Sirius 2 launched. Later dubbed Astra 5A/Sirius 2

97/12/02
Astra 1G launched from Kazakhstan. 28 transponders

97/12/19
James Cameron's 'Titanic' premiers in the USA

When TPS was launched at the end of 1997, Canalsat had 460,000 subscribers. The marketing of TPS was very aggressive and, in a few weeks, TPS signed up 100,000 subscribers. It was a complete surprise for the Canalsat management. The TPS management suddenly believed it could compete head to head with Canal+ and started to acquire some exclusive – but expensive – soccer and movie rights, with the amounts being paid exceeding that of Canal+. The Canal+ group decided to completely change its marketing policy, and it started what many saw as an aggressive period for the broadcaster.

The end result was that subscriptions to both satellite bouquets exploded, quickly catching up to cable's numbers. But Canalsat's lead increased steadily; the aggressive marketing campaign paid off. At the end of 2000, Canalsat had 1.6 million subscribers and TPS 600,000. At the end of 2004, Canalsat had 2.6 million subscribers and TPS 1.4 million.

And suddenly, in 2005, TPS and Canalsat decided to merge. The competition authorities authorised the merger. As there were more subscribers to Canalsat than to TPS, it was decided that TPS subscribers would migrate to Canalsat and Astra and, thus change their dishes. At the end of 2008, after the merger of subscribers, Canalsat emerged with an impressive 5.3 million subscribers, and cable 1.7 million. But competition is never far away.

The past few years have seen a new competitor emerge in the shape of ADSL, which uses the nation's telephony and Internet networks and has quickly achieved 6 million subscribers, half of them through Orange (the former France Télécom telephone company and now a privatised company). Satellite has one disadvantage in those urban areas, common to France and Spain, where apartments are rented. Landlords often do not permit dishes to be fixed on balconies or walls. Indeed, if the apartment does not have an easily accessible south-facing wall then they cannot 'see' the satellite. Telephony supplied ADSL (and cable) is perfect for these problem areas. However, ADSL service is limited to urban areas and thus mainly seen as a competitor to cable. Outside of these areas, and frequently even within cities and towns, satellite remains can still play on its advantages: Here, ADSL suppliers need services – and offer it with the help of satellite`s technical reach and coverage. Indeed, Orange TV's ADSL service is now marketed across France using also Astra satellites. Satellite thus has accessed a new business area and customer category: the so-called Telcos.

As a consequence, the broadcast model has now turned full circle. Satellite can play out its strengths in DTH, in direct reception, as well as in doing business with other infrastructures. Another excellent example is the introduction of digital terrestrial television in France. With the gradual switch-off of analogue signals, more terrestrial capacity has been freed up for digital transmissions. In France, as elsewhere, the Télévision Numérique Terrestre (TNT) – the terrestrial digital television signals – are broadcast from high towers scattered across the nation. As with Freeview in the UK, TNT has been on air in France since 2005 in MPEG2, and since September 2005 in MPEG4. 'Pay' and 'free' channels are available, and just as with Freesat in the UK, French viewers right across the nation are supposed to be able to access the offer. However, areas which are uncovered today by terrestrial installations can best be covered by satellite, without heavy investment in additional towers. This is how, in a curious turn, terrestrial television also comes via satellite. The offer is branded as TNT-SAT. Astra broadcasts the signals from its 19.2 degrees East craft. TNTSAT sold more than 1.8 million receivers until the beginning of 2010.

In other words satellite has a role however you slice the cake: by satellite using DTH, by satellite for services like Orange, by satellite for TNT-SAT and also by cable where satellite has always been the 'behind the scenes' supplier of most channels.

Satellite thus has accessed a new business area and customer category: the so-called Telcos.

1998

98/01/01
Smoking banned in California bars and restaurants

98/01/05
Sonny Bono, singer, dies

Not always nice to each other:
Les Guignols, a famous Canal+ favourite

98/01/21
President Bill Clinton denies affair with 'that woman'

98/02/27
UK's House of Lords ends rule of male preference

10

Countdown to Launch

10
Countdown to Launch

Geoff Bains

Although the 'Space Age' had begun with the Soviet Union's launch of Sputnik 1 in 1957, the story of rockets and man's attempts to leave the surface of the Earth goes back another 40-50 years. In the opening years of the 20th Century, scientists and academics were taking the possibilities of rocket propulsion, and transportation outside the Earth's atmosphere, seriously for the first time.

There were not that many of these dedicated rocket men at the start, and – just as it was 50 years later – most of the action took place in the USA and Russia.

In contrast to the Soviet Union's later pragmatic approach to launchers, the first red star of the space race was purely academic. Konstantin Tsiolkovsky was a self-taught school maths teacher but he laid down much of the theoretical basis for the flight of what he called 'reaction devices' which is still used today.

In the USA however, another academic, Robert Goddard, applied his theoretical findings to practical experiments. He developed practical and efficient rocket engines, proposing the use of a combustion chamber and supersonic nozzle and even developing multi-stage rockets, all essential steps to the vehicles that 75 years later would lift communications satellites into orbit.

Goddard proposed the use of liquid fuels for rockets shortly after graduating in physics and went on to build the first liquid-fuelled rocket, which he launched in Auburn, Massachusetts on March 16, 1926. This direct ancestor of modern-day launch vehicles flew for 2.5 seconds and landed in a cabbage field, but Goddard went on to improve his designs, building rockets that reached speeds approaching 900 km/h. Although he was prodigious in his theoretical and experimental work, the potential of rocketry was not recognised and Goddard was widely ridiculed; he was forced to keep much of his

Robert Hutchings Goddard (1882-1945)

98/03/25
More than 70m Astra households in Europe

98/04/29
Astra's new logo unveiled

98/03/23
70th Academy Awards sees 'Titanic' win record 11 Oscars

98/05/03
Europe agrees on the 'euro'

C. Forrester (ed.), *High Above: The Untold story of Astra, Europe's leading satellite company*, DOI 10.1007/978-3-642-12009-1_10, © Springer-Verlag Berlin Heidelberg 2011

Goddard's 1926 liquid-fuel rocket. "I wanted to ascend to Mars," he wrote as a boy. By the mid-1930s his rockets were reaching 1.7 miles. NASA's Goddard Space Flight Centre in Maryland is named after him

vision to a close circle of friends. At this time it was unusual, if not to say difficult, for non-scientists to understand that rocket propulsion operated purely by application of Isaac Newton's Third Law of Motion – that "to every action there is an opposite and equal reaction," so expelling hot combustion gasses from a rocket nozzle one way forces the rocket in the opposite direction. The public then (and even now) could understand the action of a propeller pulling a craft through air or water but a rocket was against instinctive understanding.

In reaction to Goddard's comment that rockets could take payloads beyond the atmosphere, and even to the Moon, on January 13, 1920, the 'New York Times' editorial took him to task, saying, "after the rocket quits our air and really starts on its longer journey it will neither be accelerated nor maintained by the explosion of the charges it then might have left." The anonymous editorial suggested that Goddard "does not know of the need to have something better than a vacuum against which to react."

49 years later, on July 17, 1969, the day after the Saturn V rocket left the launch pad to take Apollo 11 to the Moon, the 'New York Times' published a "correction" saying, "Further investigation and experimentation have confirmed the findings of Isaac Newton ... and it is now definitely established that a rocket can function in a vacuum as well as in an atmosphere. The Times regrets the error."

Such incidents did little to help Goddard's pursuit of rocketry excellence, nor the world's fledgling space effort. Not even the military applications of space could boost his cause. Although Goddard had developed the Bazooka anti-tank rocket grenade launcher for the U.S. Army, he was met with complete indifference when he tried to interest the US military in rockets, and so the next stage in rocket development was to take place in Europe – in Germany where the Wehrmacht, limited in long-distance weaponry by the Treaty of Versailles, turned to rockets as an alternative to artillery.

98/07/05
SES mounts its IPO. Company valued at $6 billion

98/05/11
First euro coins minted in Pessac, France

98/07/06
Hong Kong's Kai Tak Airport closed, as Chek Lap Kok airport opens

98/09/07
Google founded in Menlo Park, California

Wernher von Braun's V2

Wernher Magnus von Braun (1912-1977)

In the 1930s, Wernher von Braun joined the German military research programme and the long-range weapons developed by his design team pioneered the technology used in all launchers to this day. Of course, the best known of the rockets developed by von Braun is the 'Vergeltungswaffe 2' (Vengeance Weapon 2), or V-2 rocket used to devastating effect in World War II. The V-2 was first launched in 1942 and could rise to an altitude of 206 km. A V-2 rocket was the first man-made object to reach space, and over 6,000 of these liquid fuel rockets were built and launched in the final stages of the War.

The V-2 was just part of the Nazi rocket programme under von Braun, and other designs included multi-stage intercontinental ballistic missiles, submarine-launched missiles and the A12 that, although it never left the drawing board, let alone the launch pad, could have served to put a 10-tonne payload into low Earth orbit.

All these designs, and the V-2 in particular, set the stage for the next half century of rocket development, leading to Intercontinental Ballistic Missiles (ICBMs) in the Cold War, manned space exploration and widespread satellite deployment today.

At the close of World War II, the US and Soviet forces entering Germany in 1945, clamoured for the remnants of the Nazi rocket programme. In the end, the Soviets captured most of the unlaunched hardware while the US forces left with many of the design team, including von Braun, who was to go on to become the single most important rocket designer in the world. In effect, America had the brains and Russia the brawn.

The demonstration of rockets' military might by the V-2 during the War along with the relocation of the design team to the USA and the hardware to the Soviet Union ensured that although the development of rocket propulsion in Germany was over, rocketry was now recognised, valued and financed in the two emerging world superpowers. The space race had become part of the arms race.

98/10/01
Sky Digital TV launches in the UK

98/10/16
Chilean dictator Augusto Pinochet arrested in London

98/11/27
Sega Dreamcast is launched in Japan

An October 1945 photograph of a V2 rocket being readied for flight

98/12/10
SES announces purchase of 34% of AsiaSat

98/12/15
Eight German channels sign for extra ten years

98/12/16
Operation Desert Fox in Iraq

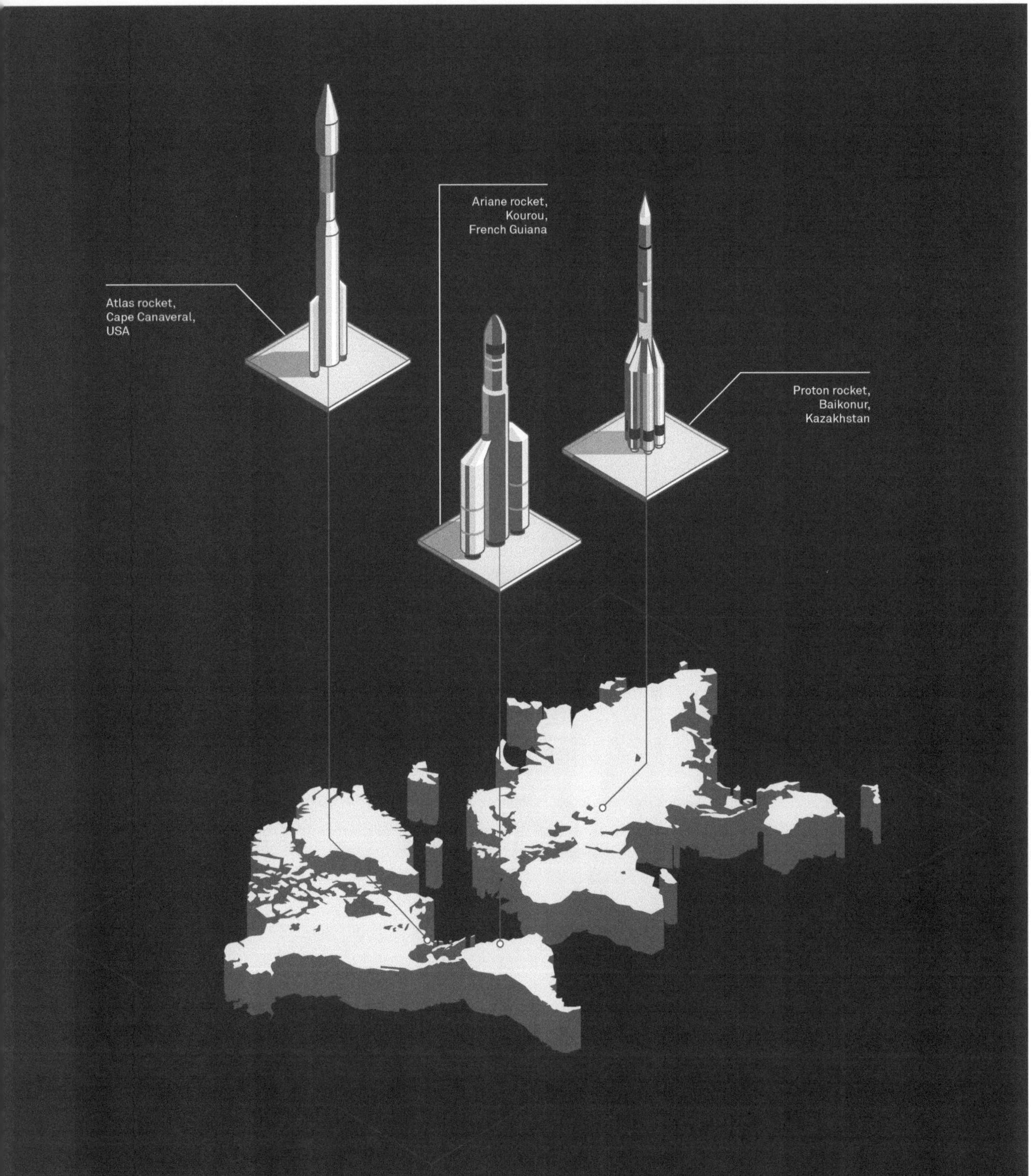

1999

99/01/07
Impeachment trial of President Bill Clinton begins

99/01/13
Basketball star Michael Jordan retires (he returns in 2001)

The United States' Atlas

Although the Soviet Union was first with a satellite and then a man in space, within a few years, the USA was ahead in launcher technology in almost every respect. In particular, the USA was launching commercial communications satellites and had the rockets to reliably put them into orbit. These rockets were originally developed for delivering, not satellites to orbit, but nuclear warheads to Soviet targets. The first was Redstone, a direct descendant of the V-2, created by von Braun's team for the U.S. Army. The Redstone missile saw service (but not active use) from 1958 to 1964. Modified versions of this rocket were used to launch the USA's first man in space, Alan Shepard, in a Mercury spacecraft on May 5, 1961, as well as to put Australia's first satellite, WRESAT, into orbit in 1967. Much of the technology used in the Redstone launcher was also later incorporated into the Saturn rockets used in the Apollo Moon programme.

Meanwhile the U.S. Air Force was developing the USA's first Intercontinental Ballistic Missile (ICBM) launcher, the Atlas. This first flew in June 1957 and became operational in 1961. A huge numbers of missiles and associated launch facilities were built but the programme was scrapped in 1966, and the surplus missiles were refurbished as launchers. Although Atlas ICBMs were never launched in anger, they proved extremely reliable as satellite launch vehicles.
The Atlas rocket used a revolutionary approach to its design, the so-called "stage-and-a-half" idea. In the 1950s it was difficult to achieve reliable ignition of liquid-fuelled engines and so single stage rockets were thought more reliable as they did not require in-flight ignition of a second stage. However, single-stage rockets required the entire weight of the large fuel tanks and engines required for lift-off to be carried throughout the flight. The Atlas launcher used a single-stage with three engines but jettisoned the outer booster engines during its flight.
What's more, the fuel tanks and body of the Atlas were little more than a stainless steel balloon held rigid for flight by the fuel pressure inside. This made the tanks extremely light (Atlas still holds the lowest dry-mass fraction of any launch vehicle ever built) so keeping empty fuel tanks un-jettisoned did not compromise the rocket's ability to carry large payloads.

Atlas launchers were used for the NASA Mercury and Gemini manned space programmes, and to launch the world's first communications satellite, SCORE on December 18, 1958. By using larger engines, extending the fuel tanks and latterly, adding solid-fuel booster rockets, the Atlas family of launchers has become ever more powerful. Ironically, for a launcher that started life as a Cold War ICBM for use against the Soviet Union, the most recent member, the Atlas V introduced in 2002, incorporates Russian-designed engines. Atlas V has been used to send planetary probes throughout the solar system and to propel satellites into orbit, including Astra 1KR on April 20, 2006.

The Soviet Union's workhorse: Proton

While the USA was developing ICBMs in the 1950s with the help of the German designers, the Soviet Union was doing the same thing using the captured V-2 rockets as its starting point. It has often been claimed that the rockets of the Russian space effort were little more than copies of the German wartime hardware. However, from the start, the Russian missiles and launch vehicles showed innovative technological advances from the team under the 'Chief Designer' Sergei Korolev who was, in effect, the Soviet's von Braun.

The first Russian rocket of note was the R-7, first flown in 1957. It was the World's first ICBM and after modification was deployed across the Soviet Union as such from 1962. However, the R-7 was not so practical as a weapon (it took 20 hours to be readied for launch, during which time the huge launch facilities required could be expected to be destroyed) and was taken out of military service in 1968.

However, the R-7 proved ideal for the commercial and scientific exploitation of space and was used for the launch of the first satellite, Sputnik 1 in October 1959, the first man in space in Vostok 1 in April 1961 and all manned flights since, including the ongoing Soyuz flights to the International Space Station.

The Soviet Union was also developing another launch system, initially as a super-ICBM. The rocket comprised a central core of first, second and third stages, and six strap-on fuel tanks and engines. It was built by Khrunichev, in the suburbs of Moscow and transported horizontally by rail to its launch site at the Baikonur Cosmodrome in Kazakhstan. First flown on July 16, 1965, the rocket carried a 'Proton' scientific satellite and that name stuck for the launch vehicle.

The Proton rocket was always far larger and more powerful than was ever required for warhead delivery, and it never flew carrying weaponry. Instead, the Proton has carried aloft a huge variety of military and scientific satellites, planetary probes, and Earth-orbiting space stations. Its initial development was also to provide the first launcher for the intended Soviet manned exploration of the Moon, but a series of failures led to the programme falling behind the US Apollo programme and the attempt was scrapped.

Alongside the Soyuz R-7, Proton became the workhorse of the Soviet, and later Russian, space effort, and it evolved into a powerful and reliable launcher. Many of the early Soviet missions, especially the failures, are shrouded in secrecy but it is believed that there have been approximately 350 Proton launches.

In 1995, the formation of a US-Russian joint venture company, International Launch Services (ILS), to operate commercial launches

First flown on July 16, 1965, the rocket carried a 'Proton' scientific satellite and that name stuck for the launch vehicle.

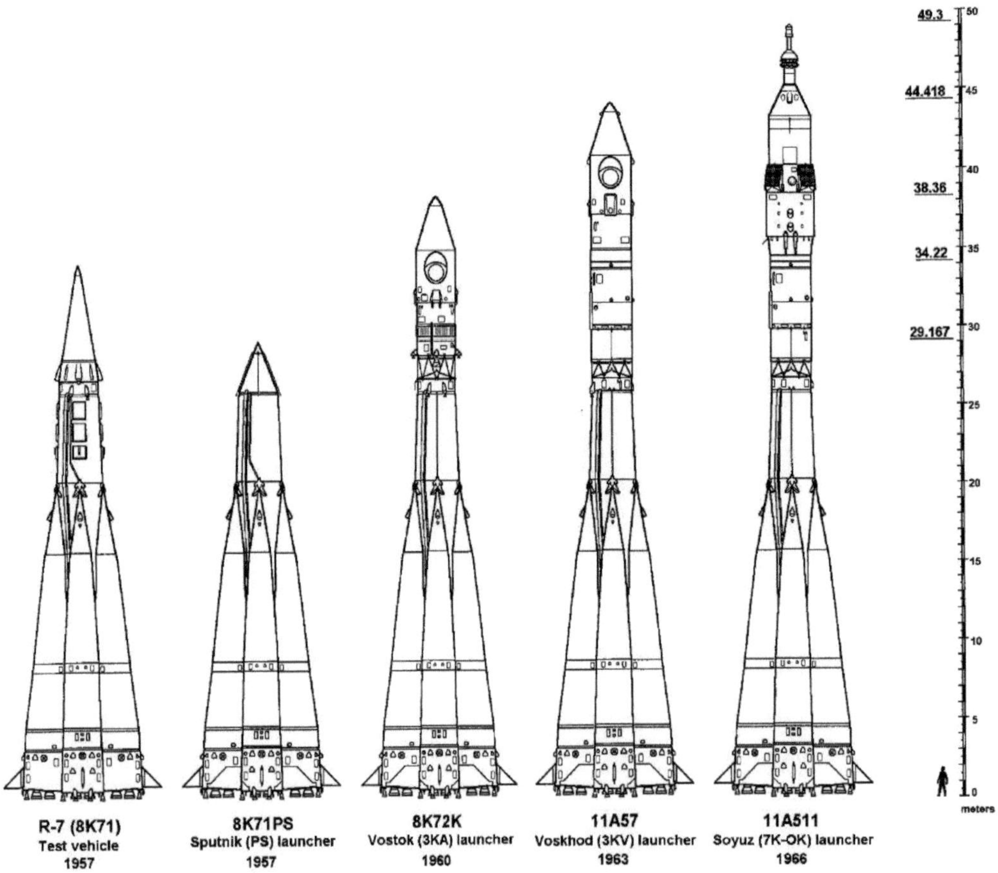

99/03/22
BSkyB takes extra nine transponders

99/03/24
Fire in the Mont Blanc Tunnel kills 39 people

99/03/24
War erupts in Kosovo

using Proton launch vehicles allowed the Proton rocket to be used for commercial satellites for the West for the first time. The first such satellite to be launched by ILS was the Astra 1F satellite on April 6, 1996 and a further seven Astra satellites have been sent into orbit from Baikonur atop Proton rockets – including, on November 26, 2002, the ill-fated Astra 1K. Astra 1K was the first (and, fortunately, so far the only) Astra satellite to be lost in a launch failure.

What flies can also fail

However, rockets do not always operate as expected. Almost every launcher ever made has had failures; sometimes these are minor problems; sometimes the launch can go catastrophically wrong, and the rocket, along with its satellite payload, is destroyed in an explosion, falls back to Earth, burns up in re-entering the atmosphere, or is stuck in a useless orbit far from where it should be (as was the case with Astra 1K).

Satellites can also develop faults. Component failures, software oversights, and even collisions with natural objects or other craft have been known. Launch disasters and in-orbit failures are not entirely avoidable. When the worst happens, there is never a repair option. The loss of a satellite can therefore only be rectified by replacement, and that takes both time and money. Building a satellite and launching it to a geostationary orbit can take several years and cost around 250 million euros, so satellite operators will usually take out insurance against loss at launch, unsuccessful deployment in orbit, and/or ongoing operational problems.

In the event of a launch failure, the launch services provider will usually provide a 'free' launch for a replacement satellite, but nevertheless, since losses are relatively frequent (launch failure rates can run to around 10%), satellite insurance premiums run to millions of euros, typically 10%-20% of the satellite cost, and the price fluctuates with the cost and complexity of the satellite, and is influenced by the recent success record of the manufacturer, the launch contractor and the satellite industry as a whole.

European efforts and Ariane's birth

While the USA and Russia were first competing against each other, and ultimately coming together to provide access to the geostationary orbit, the exodus of German expertise and hardware after the Second World War left Europe far behind.

There were some attempts at establishing European national development programmes, mainly as ICBMs for the Cold War. The UK had some success with the Blue Streak missile (at first in cooperation with the USA) launched from Woomera in Australia, and France developed its own Diamant (diamond) launcher, which notably placed the first French satellite, Astérix, in orbit on November 26, 1965 from the Hammaguir launch site in Algeria.

The two nations, together with European neighbours, pooled their efforts to create a unique European launcher, Europa. The Blue Streak missile was used as the first stage with France and Germany contributing second and third stages, and Italy, Holland and Belgium some other parts. Although the Blue Streak continued to operate well, the remainder of the rockets failed almost every test through the 1960s. British interest in the project waned and the launch site was moved from Woomera to Kourou in French Guiana.

The Europa project was abandoned in 1972 but it led to the foundation of the European Space Agency (ESA) in 1975 and the recognition that an independent launch capability for Europe was needed. So the Ariane launch vehicles were developed.

Building a satellite and launching it to a geostationary orbit can take several years and cost around 250 million euros.

99/04/20
Columbine High School massacre, 12 students and 1 teacher killed

99/05/26
First Welsh Assembly in 600 years opens in Cardiff

An Atlas 5 rocket takes off from Cape Canaveral in Florida

99/06/18
Astra 1H launched from Kazakhstan. 32 transponders, plus KaBand

99/06/13
Nelson Mandela retires from South African presidency

99/06/14
Thabo Mbeki is elected President of South Africa

Although Ariane 5 has suffered some failures, including a disastrous debut flight on June 4, 1996, together with the Ariane 4, it has become the launcher success story of the 21st Century.

The Ariane launchers were designed from the start as workhorses for lifting scientific and commercial payloads. The first Ariane 1 rocket left the Earth on Christmas Eve, 1979. Just nine years later, another Ariane rocket carried the first Astra satellite into orbit on December 11, 1988 to start the direct-to-home satellite TV era in Europe.

This was the Ariane 4 launcher, a three-stage rocket with a variety of configurations of strap-on boosters to suit different payloads. It first flew successfully on June 15, 1988, and proved a huge success, technically and commercially for 15 years. Ariane 4 was succeeded by Ariane 5, a mammoth space rocket with two stages and huge strap-on, solid-fuelled boosters, which can lift up to 10,000 kg into geostationary transfer orbit.

Although Ariane 5 has suffered some failures, including a disastrous debut flight on June 4, 1996, together with the Ariane 4, it has become the launcher success story of the 21st Century.

The Ariane rockets are launched by Arianespace, a French company formed in 1980 with small shareholdings from many European nations. Arianespace was the world's first purely commercial launch provider and has proved the most successful. To date, Arianespace has launched around 270 payloads, including more than half of the commercial satellites now operating and orbiting the planet. Arianespace now provides even Soyuz launchers – a version of the Russian R-7 rocket – for commercial launches. Arianespace launched seven of the Astra

An Astra satellite receives a final inspection at Ariane's 'clean room' at Kourou, French Guiana

99/06/25
SES listed on Frankfurt exchange

99/07/16
SES orders 12th and 13th satellites

99/07/25
Lance Armstrong wins his first Tour de France

satellites on Ariane 4 rockets and – as this is written – three Astra craft have been lifted by Ariane 5.

Key to Arianespace's success is the use of the launch site in Kourou. Close to the equator, rockets launched from Kourou pick up additional speed from the Earth's rotation, compared to launch sites such as the USA's Cape Canaveral and Russia's Baikonur Cosmodrome. This means that satellites require less fuel to reach a geostationary orbit, leaving more for their constant station-keeping adjustments in orbit, so prolonging their working life.

A satellite's long life

A satellite's long life is vital to its commercial success, and it is primarily the on-board fuel that determines a modern communications satellite's lifespan. The limited amount of fuel used to keep the satellite in position, and orientated so that its solar panels are directed towards the Sun, determines how long it can usefully serve – because, of course, there is no opportunity to refuel a satellite once it is in orbit.

The amplifiers used for the transponders' transmitters are also a potential source of problems as they can wear out after about ten years. However, most satellites are designed with extra amplifiers so as an amplifier fails or loses too much power for satisfactory use, it can be switched out of use and another switched in to replace it.

Otherwise, a satellite's solar panels keep it in electrical power almost indefinitely and it has very few 'moving parts' to wear out, although the continuous heating and cooling, in and out of the Sun's rays, can eventually damage some elements.

However, extra transmitter amplifiers or on-board fuel cannot simply be heaped on to the design regardless, as that also increases the satellite's weight, which in turn raises the launch costs. Indeed, an already heavy satellite with large amounts of extra fuel could become too heavy to launch at all.

So, satellite designers have to find a balance between the craft's weight and launch costs, and the expected life of the transmitters. The Astra 1M satellite (liftoff November 6, 2008) was designed and built to broadcast using 32 transponders, with four additional ones for use at the start of its life. Fully fuelled it weighed in at 5,344 kg, and the combination of built-in component backups and on-board fuel was expected to give the satellite a lifespan of about 15 years – a similar life to most of the Astra satellites providing Europe with television from orbit today.

The first Astra satellite was projected to have a useful life of 10-12 years. After serving for about 12 years, the craft was moved to another orbital position and put into 'inclined orbit' – a strategy that can conserve fuel but prohibits use for direct-to-home broadcasting.

At this time just nine of its original 16 transponders were operational and the Astra 1A satellite was used for sending 'feeds' and programming from one broadcasting centre to another.

Astra 1A's useful life was extended in this way for nearly three more years. Then, on December 10, 2004, Astra's first satellite was retired to a 'graveyard' orbit just one day short of the 16th anniversary of its original liftoff into the sky.

A satellite's long life is vital to its commercial success, and it is primarily the on-board fuel that determines a modern communications satellite's lifespan.

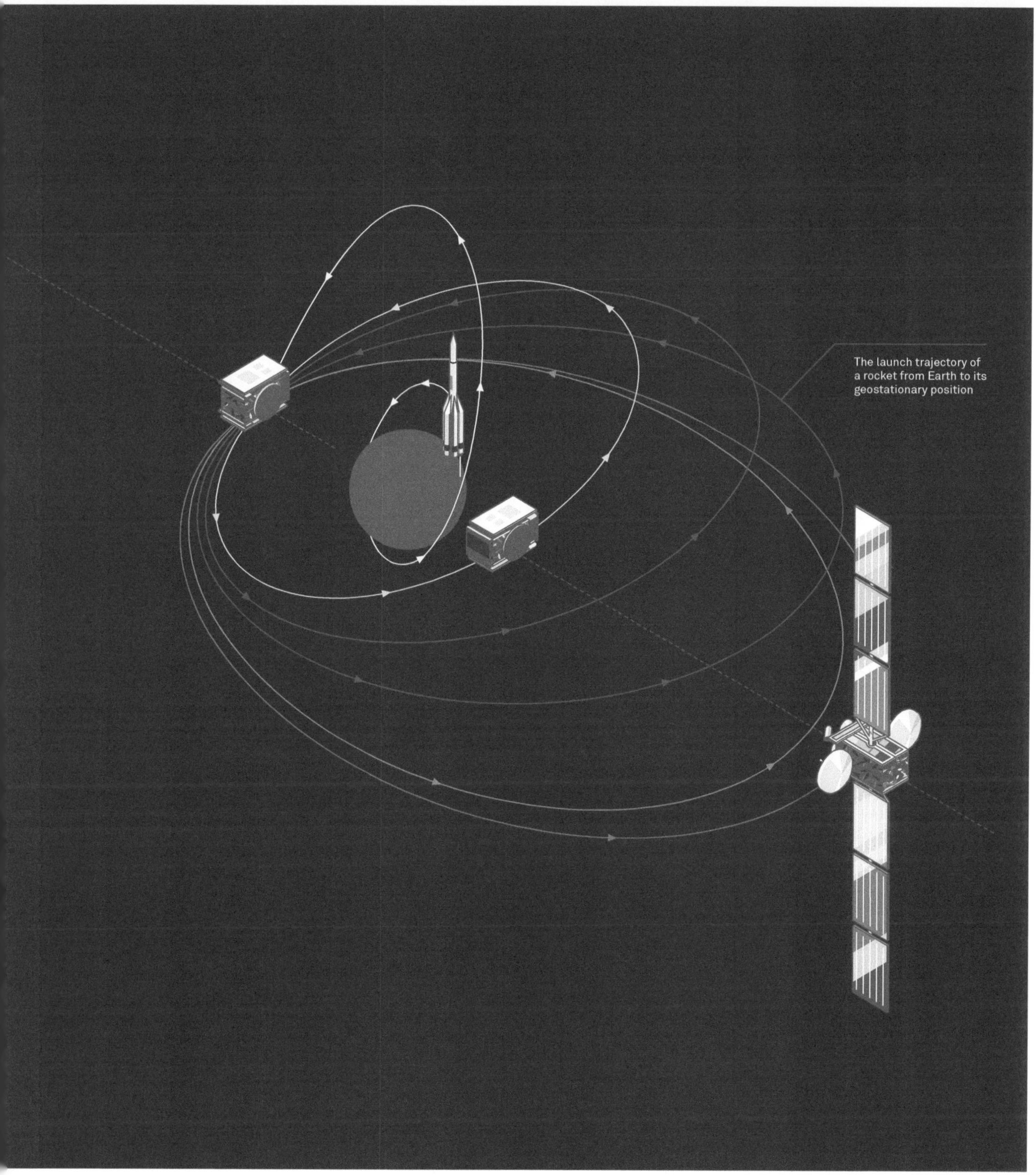

99/12/13
MTV Networks goes digital
at 19.2 degrees East

99/12/19
Desmond Llewelyn, 'Q' in 'James Bond' films, dies

11

The Brave Pioneers

11 The Brave Pioneers

Julian Clover and Chris Forrester

If we take 1950 as a generic 'starting point' for modern television broadcasting, then the medium is only now celebrating its 60th birthday. It has come a long way. To get from analogue to digital and then HDTV is not simply a technical quantum leap, it also represents the development and maturity of the analogue system. Without multi-channel analogue TV, the mass-market penetration of television sets and general economic prosperity, we would not have the necessary base ingredients to make the great leap forward.

And this great leap forward could never have happened without the inventors of television. But who were these brave men (and sorry ladies, they were all male)? The usual encyclopedia entries credit three people with 'inventing' TV. Speak to a Britisher and it is always John Logie Baird's name that will come up. But speak to a real TV historian and the response is usually Vladimir Zworykin or Philo T. Farnsworth.

The historical record is sadly devoid of references to Farnsworth. Though the oversight has begun to improve in recent years, it is still entirely possible to open an encyclopaedia and read that electronic television began when "Vladimir Zworykin invented the Iconoscope for RCA in 1923..." This sentence alone manages to express no less than three historical inaccuracies. The most conspicuous error being the "1923" date conspicuously fixes Zworykin's name chronologically before Farnsworth's 1927 patent filing, and often relegates Farnsworth to the status of 'another contributor' in the field.

Some historians have gone so far as to suggest that Farnsworth and Zworykin should be regarded as 'co-inventors'. But that conclusion ignores Zworykin's 1930 visit to Farnsworth's lab, where many witnesses heard Zworykin say "I wish that I might have invented it." Moreover, it ignores the conclusion of the patent office, in its 1935 decision in Interference number 64,027, which states

2000

00/01/01
Gisborne, New Zealand, first city to welcome new millennium

00/02/17
Microsoft releases Windows 2000

C. Forrester (ed.), *High Above: The Untold story of Astra, Europe's leading satellite company*, DOI 10.1007/978-3-642-12009-1_11, © Springer-Verlag Berlin Heidelberg 2011

quite clearly "priority of invention awarded to Farnsworth."

These misinterpretations of the historical record are precisely what more than eighty years of corporate public relations wants us to believe – that television was "too complex to be invented by a single individual." But close examination of the stories beneath the written record reveals a far more compelling story: In fact, there was one inventor of electronic television. Video, as we now know it, first took root in the mind of Philo T. Farnsworth when he was just fourteen years old, and he was the first to successfully demonstrate the principle, in his lab in San Francisco on September 7, 1927. If you need to fix a date on which television was invented, that's the date.

Before that date, television was the province of Newtonian electro-mechanical engineers who employed spinning disks and mirrors in their crude attempts to scan, transmit, and reassemble a moving image. The inventions of Jenkins, Ives, Alexanderson, Baird,

A woman's face as received on 30-line mechanical television

00/03/13
SES included in FTSE 'EuroTop 300' index

00/03/26
Vladimir Putin elected
President of Russia

00/05/04
Ken Livingstone is first-ever
Mayor of London

United States - 1939: TV pioneer Philo Farnsworth with early TV equipment

00/07/27
SES buys 20% stake in Brazil's Embratel

00/06/19
Tiger Woods wins US Open by 15 shots, a record

00/07/25
Concorde crashes on takeoff, kills all 109

Philo T. Farnsworth (1906–1971)

and others are all similar in their reliance on the spiral-perforated, spinning disk first proposed in the 1880s by the German Paul Nipkow. These contraptions were engineering marvels in their own quaint way, but they were not the sort of breakthrough that Farnsworth introduced, nor is anything left of their technology in the system of television that is in use around the world today.

On September 7, 1927 Farnsworth demonstrated for the first time that it was possible to transmit an "electrical image" without the use of any mechanical contrivances whatsoever. In one of the first triumphs of Relativistic science, Farnsworth replaced the spinning disks and mirrors with the electron itself, an object so small and light that it could be deflected back and forth within a vacuum tube tens of thousands of times per second. Farnsworth was the first to form and manipulate an electron beam, and that accomplishment represents a quantum leap in human knowledge that is still in use today. After September 7, 1927, every new contribution to the art – including Zworykin's – was an improvement on Farnsworth's simple, elegant, and profound invention.

What is so often overlooked cannot be overstated: In 1923, Vladimir Zworykin – recently emigrated from Russia, and employed at the time by the Westinghouse Corp. in Pittsburgh – applied for a patent for an approach to television that he first encountered in the classroom of Boris Rosing, his former teacher in Russia. In 1927, Farnsworth also applied for a patent. Later that year, Farnsworth produced the first successful transmission of a television image by wholly electronic means – an event that is thoroughly documented in Farnsworth's journals – while Zworykin's application was still pending. Farnsworth's patent, number 1,773,980 – with its critical Claim 15 regarding the "electrical image" – was issued in August 1930 – while Zworykin's application was still pending.

The 1923 Zworykin application would be forgotten – except that a patent for the Iconoscope was finally issued in 1938 bearing a 1923 application date. This patent (number 2,141,059) was issued an extraordinary fifteen years after the application date, and then only after extensive revisions had been made to the original application.

Furthermore, the eventual patent granted pursuant to the 1923 application was issued over the objection of the patent office, and even then not until the case was adjudicated by a court of appeals. That the Iconoscope patent was issued at all hinged on a technicality, and it served no practical purpose other than substantiating the dates that RCA would eventually use in its public relations campaign.

RCA's obtaining of the patent in 1938 has served as the cornerstone of its efforts to influence the historical record, since the patent effectively fixes 1923 as the date that Zworykin first disclosed electronic television. Decades later, historians and scholars are still including this dubious 1923 date in their chronologies.

What's wrong with the Zworykin patent? What's wrong with it is that the original application – the system that Zworykin disclosed in 1923 – simply could not work. The idea was on the right track, but the application fell far short of disclosing a device which would pave the way to electronic video and ultimately put a television in every living room or a computer monitor on every desktop.

Some historians suggest that witnesses observed some sort of blurry smudge. Zworykin would claim years later that the image of a cross was transmitted.

Vladimir Zworykin at RCA

There is scant evidence that Zworykin ever built and tested a system like the one disclosed in his 1923 application. One story does exist about Zworykin's attempt to demonstrate his concept for executives of Westinghouse, where he was employed at the time, in hopes of obtaining more funding for his research. The demonstration was so dismal that instead of providing him with further funding, Zworykin's superiors ordered him to find something "more useful" to work on.

The usual retelling of this story is cast in such a way that we are supposed to believe that the Westinghouse executives who witnessed and dismissed this demonstration were too shortsighted to appreciate its promise. It seems more plausible to conclude that what they saw showed little promise because it simply didn't work. Some historians suggest that witnesses observed some sort of blurry smudge. Zworykin would claim years later that the image of a cross was transmitted. But during the critical 1934 interference proceedings there was no evidence submitted to support even these modest contentions.

Historians should focus more carefully on the decision of the U.S. Patent Office in its historic 1935 ruling in Patent Interference Number 64,027. This is the litigation in which Zworykin challenged Claim 15 in Farnsworth's patent number 1,773,980, which describes the "electrical image". An electrical image is the electrical counterpart to an optical image. When an optical image is focused on a photoelectric surface, the light-sensitive chemicals emit an array of electrons – the "electrical image" – which can then be scanned to form a fluctuating current. That is the very essence of how an electronic television signal is created, and so it is understandable that Zworykin and RCA would attempt to appropriate the language in this claim. There is simply no getting around it – you can't create an electronic television signal without first creating an "electrical image".

The whole of RCA's research effort – at an expense that David Sarnoff joked with Zworykin years later cost RCA more than 50 million dollars – was intended to circumvent Farnsworth's patents, in particular Claim 15. When the electrical image in Claim 15 proved essential, Sarnoff, Zworykin, and RCA's attorneys went to great lengths in the 1934 interference to prove that the 1923 application would have created such an electrical image, and that Zworykin was therefore entitled to 'make the count' embodied in Claim 15.

But when it was time for RCA to produce material evidence that Zworykin had constructed and operated his system in 1923, there was no evidence submitted. No tubes were displayed, no laboratory journals entered into the record. There were only confusing and contradictory verbal accounts from two Zworykin colleagues.

00/09/14: Astra 2B launched from French Guiana, serves UK and Ireland

00/10
SES buys 50% of Nordic Satellite Co (NSAB)

00/09/15
2000 Summer Olympics open, Sydney, Australia

00/10/30
Soyuz TM-31 launches, taking first crew to International Space Station

00/11/07
George W. Bush defeats Al Gore in US elections

After considering all the testimony, the patents examiners ruled in Interference number 64,027 that "Zworykin has no right to make the count because it is not apparent that the device would operate to produce a scanned electrical image unless it has discrete globules capable of producing discrete space charges and the Zworykin application as filed does not disclose such a device."

The patent examiners were unequivocal in their decision to award the indispensable Claim 15 to Farnsworth. The case was appealed and RCA lost all the appeals. This pattern went on, over this and other patents, until RCA capitulated in 1939 and accepted a license from Farnsworth for the use of his patents – the first such license in the history of a company that was determined to "collect patent royalties, not pay them."

There is simply no getting around it – you can't create an electronic television signal without first creating an "electrical image".

The authors gratefully acknowledge use of text from The Farnsworth Chronicles (www.farnovision.com) in compiling the early historical background to 'The Brave Pioneers'.

A staged boxing match being filmed at television studios at Crystal Palace.

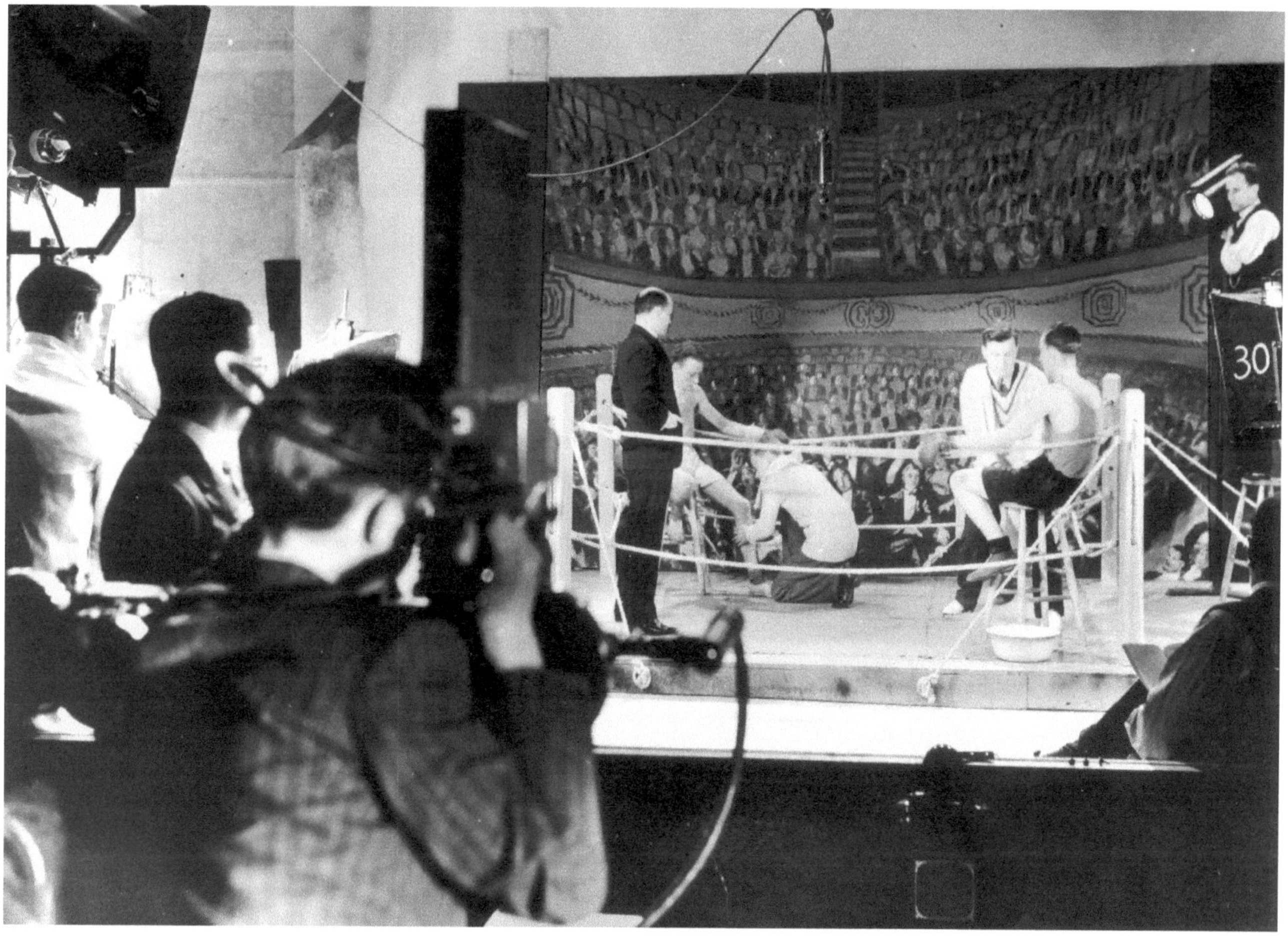

00/12/18
Astra buys 10% stake in ND-Satcom

0/12/19
Astra 2D launched, expanding UK channels

00/12/15
Chernobyl Nuclear Power Plant is fully shut down

Rupert Murdoch

The beginnings of Rupert Murdoch's association with satellite television came in 1983 with the purchase of Sky Television, a small UK-based operation that had been launched by the former Thames Television researcher Brian Haynes. With a schedule that included a Dutch-accented puppet by the name of DJ Kat and the Australian technology show "Beyond 2000", Sky was beginning to establish itself in the more developed cable markets of the Benelux and Scandinavia. Murdoch wanted more and sent a team to the Isle of Man, first to convince the authorities that launching their own TV channel was a viable project – there was more local interest in creating a new Radio Luxembourg-style operation – and to do the tests to see how well the signal travelled into the area of Northwest Britain affectionately known as Granadaland.

But why target a few thousand people when a change of satellite could bring you many millions? While British Satellite Broadcasting jumped through the hoops put in place by the then UK regulator, the Independent Broadcasting Authority, Murdoch broke bread with newly established satellite operator Astra. In the spring of 1988, Murdoch signed his four-transponder agreement, announcing his coup to a packed dinner of BAFTA, the British Academy of Film and Television Arts, the British 'Oscars'. The four-channel package consisted of the then Sky Channel (later Sky One), Sky News, Sky Movies and Eurosport, the latter a joint venture with the European Broadcasting Union. After the launch in February 1989, Sky Channel remained on the Eutelsat satellite at 13 degrees East for but a few months before its pan-European distribution was handed to Eurosport. When Sky later merged with BSB, The Sports Channel became the grounding for Sky Sports, Eurosport came close to collapse before being rescued by the French commercial channel TF1.

There were two other channels slated at launch. Sky Arts, and The Disney Channel, both would eventually launch though not until much water had flowed under the bridge. A newspaperman through and through, Murdoch was happy to maintain the loss-making Sky News, because of the credibility it brought to the operation. There may have been cheap imports on Sky One, and out of primetime they were seriously cheap, but Sky News was the perfect riposte. Murdoch would also use some of his chief newspaper lieutenants, first the blunt British journalist Andrew Neil, who was the successful editor of 'The Sunday Times' as Sky Television's Executive Chairman. In January 1994, Kelvin MacKenzie, the opinionated former editor of 'The Sun', became Managing Director of Sky, but this appointment was to last only a few months. Another newspaperman, Vic Wakeling, ran Sky Sports very successfully for almost two decades until his retirement in June 2009.

The merger with BSB, which could just as easily be described as a takeover, was to continue to weigh heavily on Rupert Murdoch and News Corp. In January 1991, barely a month after the merger, 8.2 billion dollars of debt was restructured. Somehow Murdoch knows the right moment to gamble.

Leo Kirch

One of the most colourful figures in German broadcasting, Leo Kirch controlled the holding company ProSiebenSat.1 until his KirchMedia entered into well-publicised bankruptcy in 2002. Up to that time, Kirch also owned the Premiere pay-TV platform, both of which were key movers in putting German television onto Astra.

As the founder and main shareholder of the KirchGroup, Kirch began his activities with a holding firm that bought up mainly US TV rights and then sold them on to the public broadcasters. With a grand vision, he started to build a broadcasting empire that would control all the stages of exploitation: buying up the rights, then bringing them to premium pay TV and later on to free TV.

However, the premium business proved to be fatal, and it was also the losses incurred at Premiere that led to the company's bankruptcy in 2002. The Premiere brand was consigned to history in July 2009, following its purchase by News Corp, and rebranding as Sky Deutschland. Leo Kirch launched its predecessor, Teleclub, as the first German pay-TV operation, in Switzerland at the beginning of 1988. Two years later, KirchGroup, UFA and Canal+ went on to form Premiere, with the service officially making its debut on February 28, 1991.

Teleclub has since been screened only in Switzerland, its German subscribers having been integrated into Premiere. On July 28, 1996, DF1 was launched as the first digital pay-TV offering in Germany, with the KirchGroup as its major shareholder. Building a complete platform, Kirch chose to form a vertically integrated business. The technical company BetaTechnik meanwhile developed the so-called d-Box with its own encryption system, known as BetaCrypt.

When the KirchGroup filed for bankruptcy in 2002, it was largely the financial burden of the pay-TV division that led to the financial debacle. In a way, by pioneering digital television, the company was also responsible for making consumers associate digital with pay TV, and this was certainly an impediment with the introduction of digital technology in the country.

01/02/01
Astra 2D operational

01/03/15
Astra's reach tops 87 million European homes

01/02/13
El Salvador 6.6 earthquake killing 400

01/03/23
Russia's Mir space station de-orbited near Fiji

Did you own one of these?

1929

1970

1950

1956

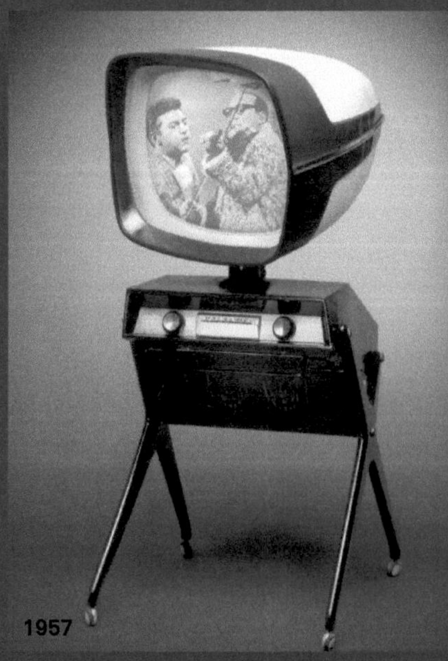

1957

1941

1947

1943

1958

1972

1960

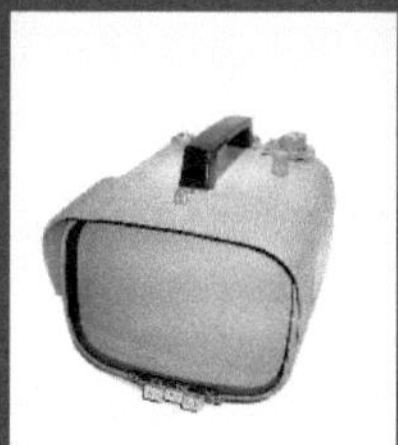
1661

1972

1978

1960

2010

1963

1959

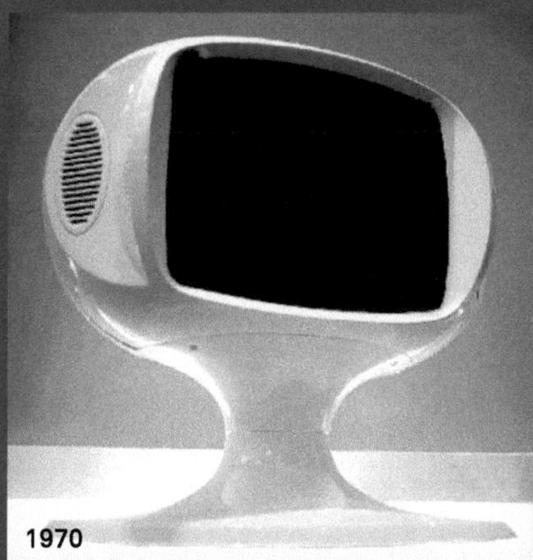
1970

1965

12

Satellites: The 'Early Birds'

12
Satellites: The 'Early Birds'

Geoff Bains

When the Astra 1A satellite left the launch pad at Kourou in French Guiana on December 11, 1988 to kick off the era of direct-to-home satellite television in Europe, it was 30 years nearly to the day after the very first communications satellite took off for space.

Project SCORE (Signal Communications by Orbital Relay Equipment) was a far cry from the TV satellites of today, but it left the launch pad in Cape Canaveral, USA on December 18, 1958 to become the first satellite to receive and re-broadcast radio signals, and for its 12-day life, it clearly demonstrated the practical possibilities of communications satellites.

Surprisingly, the SCORE satellite was very much on par with today's giants in terms of size. It was one of the largest satellites ever built – because, rather than orbit as a separate unit, the payload was actually built into the USAF Atlas ICBM rocket that launched it, and the whole rocket (bar the engines) orbited the Earth.

Like the Astra satellites, SCORE was designed to survive faults, and it contained two receive and retransmit transponders and two tape recorders to provide a 'record-and-forward' mode – essential as the craft circled the Earth in a low elliptical orbit in just over 100 minutes, so it was only accessible for short periods and tracking for both uplink and downlink was difficult.

Quite unlike the extensive planning and preparation that goes into each modern satellite, the complete specification, construction and launch of SCORE remarkably took just six months. The backup strategy proved wise as problems developed with the first transponder, and the second was brought into use for the numerous real-time and recorded-and-retransmitted radio relays.

01/04/02
SES buys GE Americom, for $5 billion, expands into North America

01/05/28
Soyuz TM-32 launched carrying first space tourist, Dennis Tito

01/06/06
Astra 2C is launched. 32 new transponders

01/06/07
Tony Blair wins UK election

C. Forrester (ed.), *High Above: The Untold story of Astra, Europe's leading satellite company*,
DOI 10.1007/978-3-642-12009-1_12,

One of the recorders was to be launched with a pre-recorded message from the satellite assembly team but at the last minute, as the rocket stood on the launch site, it was replaced with a Christmas goodwill address from US President Eisenhower, and this was broadcast to the world from space on December 19, 1958.

The SCORE satellite was little more than an experiment (and something of a political statement by the USA, a year after the USSR had surprised the world with the first-ever man-made satellite, Sputnik 1) but it clearly demonstrated the possibilities of communications satellites and paved the way for their development right up to the sophisticated and powerful spacecraft in use today.

A Christmas goodwill address from US President Eisenhower was broadcast to the world from space on December 19, 1958.

The next step towards satellite TV was the first satellite to relay television pictures, but in some ways, this was a bit of a retrograde step for satellite development. Echo 1, launched by NASA on August 12, 1960 was only a passive communications satellite.

It was simply a spherical balloon 30 metres wide and made of a metallised plastic film so that radio waves broadcast from the ground would reflect off it, back to Earth. This simple system had the advantage that

01/06/11
Oklahoma bomber Timothy McVeigh executed

01/06/21
Slobodan Milosevic, former Yugoslav President, arrives at The Hague for trial

One of NASA's experimental balloon reflecting satellites

A Telstar engineer preparing the protective faring for the satellite

there was little limit to the frequencies or bandwidth of the signals that could be used, and multiple signals could be handled at once. What's more, it was relatively cheap to build and to launch, and the satellite was used to test the point-to-point transmission of voice, pictures and data. Like SCORE, Echo 1 was launched into a low, elliptical orbit and was usable only for a few minutes at a time. Nevertheless, several successful transmissions across the USA were made and even some across the Atlantic, to the UK.

Since there was practically nothing to go wrong in the satellite, not even batteries to run out of power, Echo 1 lasted over eight years before its orbit decayed.

Although NASA launched another, larger passive satellite, Echo 2, on January 25, 1964, which was also used for both communications and scientific experiments, this was the last 'sateloon' produced, and the idea of passive communications satellites was abandoned for active satellites that could amplify and retransmit television signals.

The first of these was also the first satellite sensation. Telstar was a worldwide publicity phenomenon, with countries around the globe clamouring to use, or just be associated with, the new technological wonder. It even inspired an international hit record from The Tornados.

The idea of passive communications satellites was abandoned for active satellites that could amplify and retransmit television signals.

Telstar 1 pioneered many of the technologies still used in the television satellites of today, such as TWTA (travelling wave tube amplifier) transponders and electrical power provided by an array of solar cells.

Telstar was built by AT&T's Bell Telephone Laboratories as the experimental prototype for a complex 500 million-dollar system

01/07/23
Romain Bausch appointed President and CEO of SES Global

01/08/22
Astra 1D operational

01/07/23
Kyoto Protocol signed by 178 nations

of 55 low orbit communications satellites intended to establish a private US monopoly over international telephone and TV communications. Of course, the AT&T satellite system never reached fruition, but a complex system of communications satellites was built by many companies and countries at a cost many times this.

First 'live by satellite' TV connection began with Kennedy saying, "I understand that part of today's press conference is being relayed by the Telstar communications satellite to viewers across the Atlantic."

First TV signals to cross the Atlantic in 1962 from the tiny Telstar1 satellite

The satellite was tiny by today's standards – a sphere about 90 cm in diameter and weighing just 77 kg. The receive and transmit antennas were in a belt around the satellite's middle with most of the rest of the aluminium shell covered in the power-providing solar cells. Inside was suspended the electronics package, including a single TWT amplifier transmitting 3W of power. The whole Telstar 1 project is estimated to have cost AT&T about 70 million dollars.

Telstar 1 was launched from Cape Canaveral on July 10, 1962 and it began operations the same day. The US and French antennas were horn designs, but at Goonhilly Downs the signals were received on the world's first open parabolic satellite dish, forerunner of the many millions of small domestic receiving dishes now in use across Europe – although at 26 metres in diameter, Goonhilly's 'Arthur' dish was a little larger; even bigger than today's uplink dishes at the Astra headquarters in Luxembourg.

The first television signal to cross the Atlantic, just 15 hours after launch, was of a black and white recording of a US flag fluttering in front of the Andover station with the US national anthem as a soundtrack. The flagpole was actually held out of shot by a technician and the 'wind' provided by a fan!

Test broadcasts of recorded television, pictures and telex were sent in both directions across the Atlantic for two weeks and then the first live TV transmission was made – of a press conference by US President Kennedy on July 23.

After a few minutes of a televised baseball game, this first 'live by satellite' TV connection began with Kennedy saying, "I understand that part of today's press conference is being relayed by the Telstar communications satellite to viewers across the Atlantic, and this is another indication of the extraordinary world in which we live."

Extraordinary it was, but ironically for a technological development greeted across the globe as a beacon for international cooperation and peace in the Cold War world, Telstar was prematurely silenced by radiation from a US high-altitude nuclear test conducted the day before its launch, and it failed in December 1962. The ground team managed to restart Telstar in January 1963, but further radiation damage killed the satellite for good a month later.

Although completely inactive, Telstar 1 is still in orbit around the Earth today. Telstar's performance was a resounding success, but it proved the problems of low elliptical orbits for communications satellites. Telstar's television feeds could last for only about 20 minutes of the satellite's 158-minute orbit, and the future for global communications was clearly with geostationary satellites.

That torch was taken up by the Hughes Aircraft Company (later Hughes Space & Communications and now Boeing Satellite Systems, and builder of some of the Astra satellites in orbit today) with the Syncom range of satellites built for NASA. Syncom I failed on launch and Syncom II operated in a geosynchronous orbit and not a geostationary one (so it appeared to move north and south during each day) but Syncom III was launched on August 19, 1964 and became the first communications satellite to use the geostationary orbit, so it seemed to remain in one spot in the sky.

01/09/14
SES demonstrates MPEG4 transmission at IBC

01/09/11
September 11 attacks on World Trade Center, and elsewhere in US

1964: Former US Olympic champion Rafer Johnson discusses the Syncom III communications satellite with Chinese Olympic decathlon contestant C K Yang, Tokyo, Japan

01/10/07
Tora Bora bombings against Taliban begin

01/10/23
Apple unveils iPod

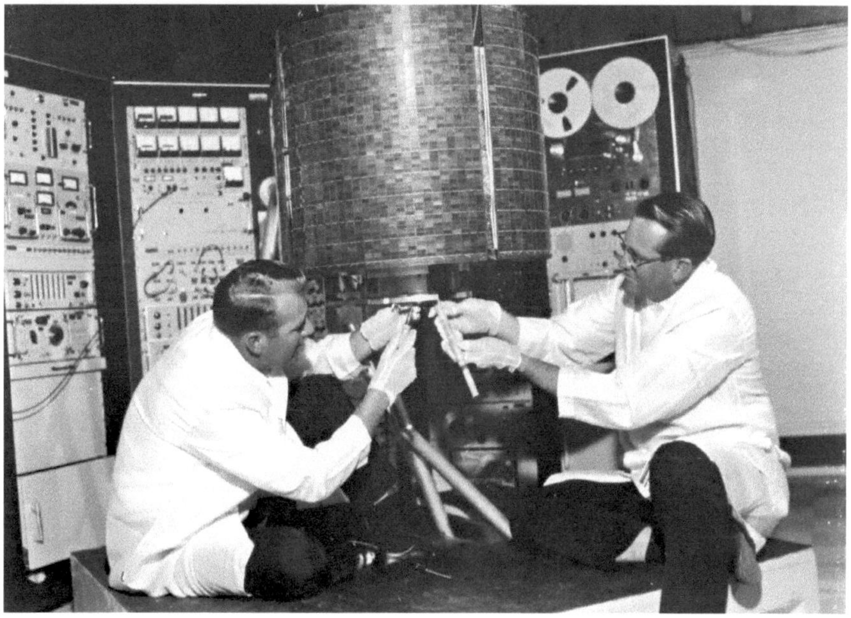

Technicians working on Early Bird

In US magazine, 'Popular Science', this new geostationary satellite was nicknamed "Stay-putnik"!

Syncom III was a cylindrical craft, the width of an oil drum (but about a third the height) with a covering in solar cells and a dry weight of just 25 kg. It was stabilised in space by spinning the body (as are the Astra 2D and Astra 3A satellites, built by Hughes and descended from this design) and, like SCORE, it also carried two receivers and two 2W transmitters in case of failure.

Some of the details of Syncom III's performance remain shrouded in secrecy because of its use by the US military (the satellite played a key role in communications in the Vietnam conflict). However, a month after its launch, Syncom III very publicly provided the USA with live coverage of the Summer Olympic Games in Tokyo, the first continuous television to cross the Pacific Ocean.

The Syncom III satellite was only expected to last for 12 months but it remained in service until April 1969, by which time its successor had taken geostationary communications satellites away from the experimental and the military, and pushed them firmly into the commercial domain.

Early Bird was also built by Hughes and launched on April 6, 1965 for Comsat, a private company created by the U.S. Congress especially to develop commercial satellite communications. The craft was a little larger than the Syncom satellites and able to receive and retransmit 240 telephone circuits or one TV channel across the Atlantic between Europe and the Americas.

Early Bird is rightfully one of the best-known pioneering satellites, not least because the public's imagination was then caught by the same power and versatility of satellite television that Astra employs to this day.
In its four years of service, Early Bird made thousands of transatlantic connections but it also made headlines when heart surgery in Houston, USA was observed by doctors in

In its four years of service, Early Bird made thousands of transatlantic connections but it also made headlines when heart surgery in Houston, USA was observed by doctors in Geneva, Switzerland.

01/11/12
SES completes GE Americom purchase

01/11/12
American Airlines Flight 587 crashes at JFK airport. 260 die

01/12/15
Leaning Tower of Pisa reopens after 11 years

Geneva, Switzerland; when European viewers saw live pictures of troops in the streets of Santo Domingo, Dominican Republic; when British viewers watched a live telecast of the Cassius Clay vs. Sonny Liston world heavyweight title fight in Maine, USA; and when Canadian bank robber, Georges Lemay was recognised in Florida, USA after his mug shot had been shown on TV.

Such feats are commonplace today but in the 1960s, Early Bird's operation seemed nothing short of miraculous. Perhaps the potential for satellite TV, and the future of television, with millions watching television in their homes via satellites, was most clearly glimpsed with the 'Our World' television programme in 1967 – when Maria Callas, Pablo Picasso, and The Beatles were among performers contributing to an international production broadcast live from 14 countries by three satellites including Early Bird to 31 countries worldwide and an estimated audience of 400 million.

Impressive though satellite TV was in the 1960s, it was still providing only a point-to-point connection between broadcast centres. Transmission direct to viewers' homes had yet to arrive, for that needed further technological advances.

NASA's ATS-6 satellite was the first to be used to distribute TV signals to domestic viewers. ATS-6 was launched on May 30, 1974 and it carried the SITE (Satellite Instructional Television Experiment) project to bring educational TV programmes to remote communities in India.

This was also the first geostationary satellite to use '3-axis stabilisation' – tiny thrusters maintained the crafts attitude to the Earth – instead of the spin stabilisation of Syncom and Early Bird. Most of the Astra satellites use 3-axis stabilisation.

The SITE project ran for one year from August 1975 with educational programming uplinked from India to ATS-6 and retransmitted using a 20W transponder at UHF frequencies to 2,000 rural villages equipped with community television receivers

The 1967 'Our World' transmission featured the Beatles, then at the height of their fame

and 3-metre dishes (so this wasn't strictly direct-to-home TV). SITE not only demonstrated the benefits of education to rural communities but also proved the viability of television distribution by satellite.

The Russian Ekran satellites were the Soviet Union's first geostationary satellites, and also the first in the world to broadcast TV direct to individual homes. Launched in the 1970s, the Ekran satellites were largely for

The Russian Ekran satellites were the Soviet Union's first geostationary satellites, and also the first in the world to broadcast TV direct to individual homes in the 1970s.

experimental purposes but provided over 18 million Soviet citizens access to central television from Moscow.

The first satellites carried just one colour TV channel and this was transmitted in UHF frequencies for reception with a normal TV aerial, but that required a 200W transmitter – a transmission power that's difficult to match even by today's satellites (largely because they carry multiple transponders).

2002

02/01/09
Michael Jackson receives Artist of the Century award

02/02/01
'Wall Street Journal' reporter Daniel Pearl murdered in Pakistan

The problem of transponder power is simply where to get the power from. The earliest satellites used an on-board battery and this determined the life of the satellite. Solar cells provided electrical power for later craft but as the power and number of the transponders increased so the power demands of the whole satellite grew too, and large areas of solar cells were needed.

The single 200W transponder on the Ekran satellites required 25 square metres of solar cells to power it. The latest Astra satellites have 32 or more transponders, each of 150W, so although modern solar cells are vastly more efficient, the whole satellite still requires solar panels over 70 square metres in size, providing more than 9kW of power – enough to power several homes back on Earth.

A further difficulty is that the satellites are not in sunlight all of the time, and when their orbit takes them into the shadow of the Earth, the solar cells cannot provide power. So satellites are again equipped with batteries, which are recharged by the solar cells during the satellite's 'day' to power the craft during its 'night'.

This 'eclipse protection' is essential for TV satellites. Quite apart from many channels requiring 24-hour transmission, depending on their orbital position, the satellite's hours in darkness may not correspond to the night time in the area of the Earth receiving the TV transmissions. Every Astra satellite has full eclipse protection for all of its transponders so it can keep broadcasting television channels to the viewers below spread across Europe, all day and all night.

In the years following SITE and Ekran, satellites were increasingly used to deliver TV signals from central facilities to many local distribution centres, where the signals would be taken to individual homes by terrestrial transmission or by cable.

In the USA in particular, the need for local cable networks to obtain their programming and to provide the three national TV networks (CBS, NBC, and ABC) with a means to deliver programming to their local affiliates, led to the rise of satellite distribution.

RCA's Satcom 1 (launched December 13, 1975) was the first satellite to be used by the three networks and was also the birthplace of satellite-delivered cable channels such as HBO and CNN. A key reason for Satcom's commercial success in the face of competition from other satellites for TV distribution business was its capacity – Satcom 1 provided 24 transponders so the cost per transponder to broadcasters was reduced.

Capacity is still a key attribute for TV satellites as ever more channels use the satellite route to viewers, and Astra has both initiated and met the huge demand of broadcasters to reach the eager TV viewers across Europe, using a combination of established and innovative techniques.

Polarising the broadcast signals in both horizontal and vertical planes increases the number of transponders that can be 'fitted in' to the frequencies available. Intelsat IVA F1, launched in September 1975, was the first craft to transmit dual-polarised signals and today all Astra satellites use this technique to provide the maximum capacity to customers from each satellite.

Of course, more customers can also be satisfied by transmission from different orbital positions – so a satellite at one orbital position can use the same frequencies as another close by, provided the positions are far enough apart so that dishes on the ground can distinguish between the satellites. We examine these developments in Chapter 13 'The Satellite Miracles'.

02/02/15
Romain Bausch is declared 'Satellite Executive of the Year'

02/03/15
Astra's German coverage tops 39 million homes

02/03/20
Astra reach confirmed at more than 91 million homes

02/03/27
Dudley Moore, British actor, dies

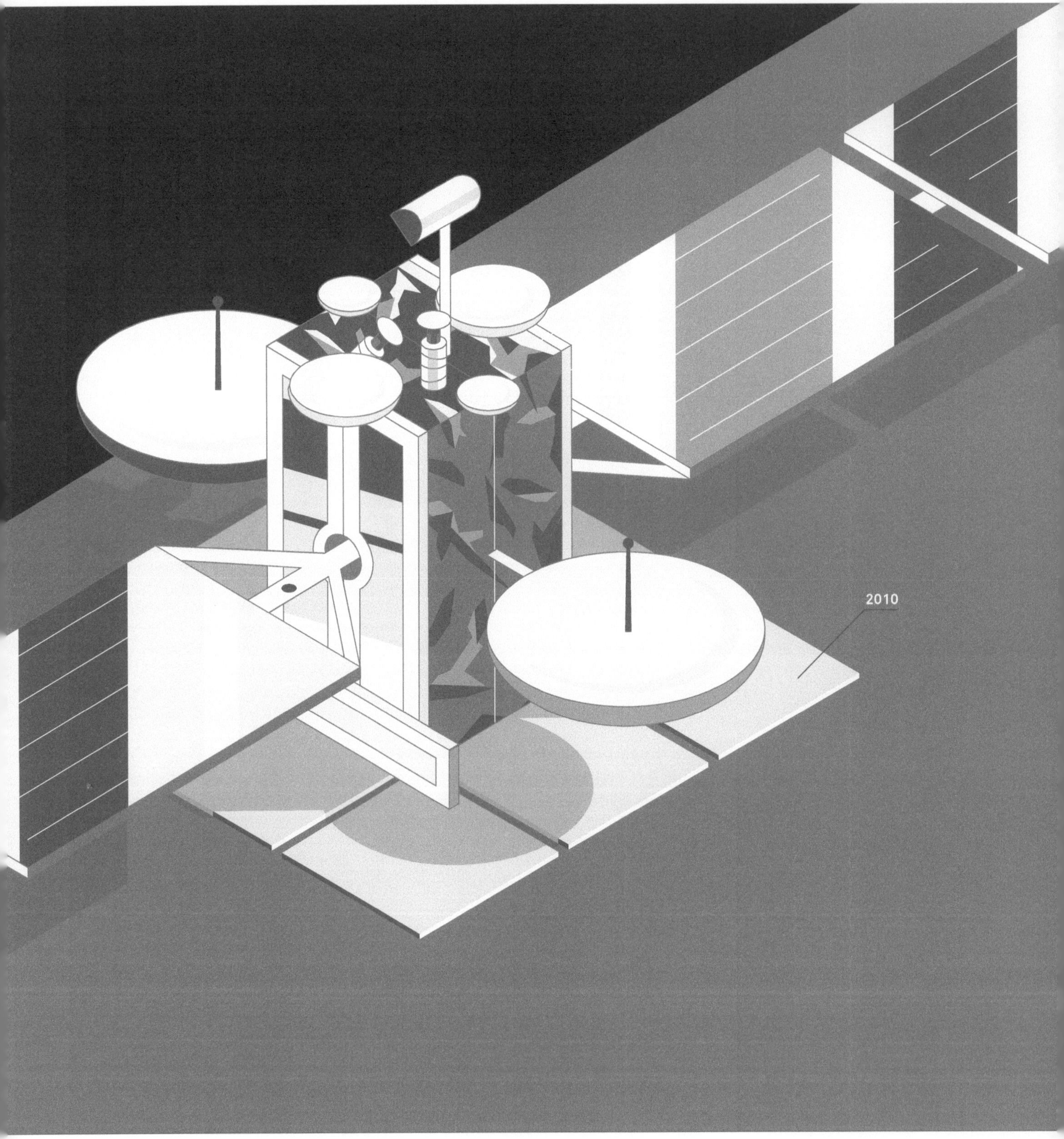

02/03/29
Astra 3A is launched. New orbital position at 23.5 degrees East

02/04/30
Astra 3A operational

02/04/09
Funeral of Queen Elizabeth, the Queen Mother, at Westminster Abbey

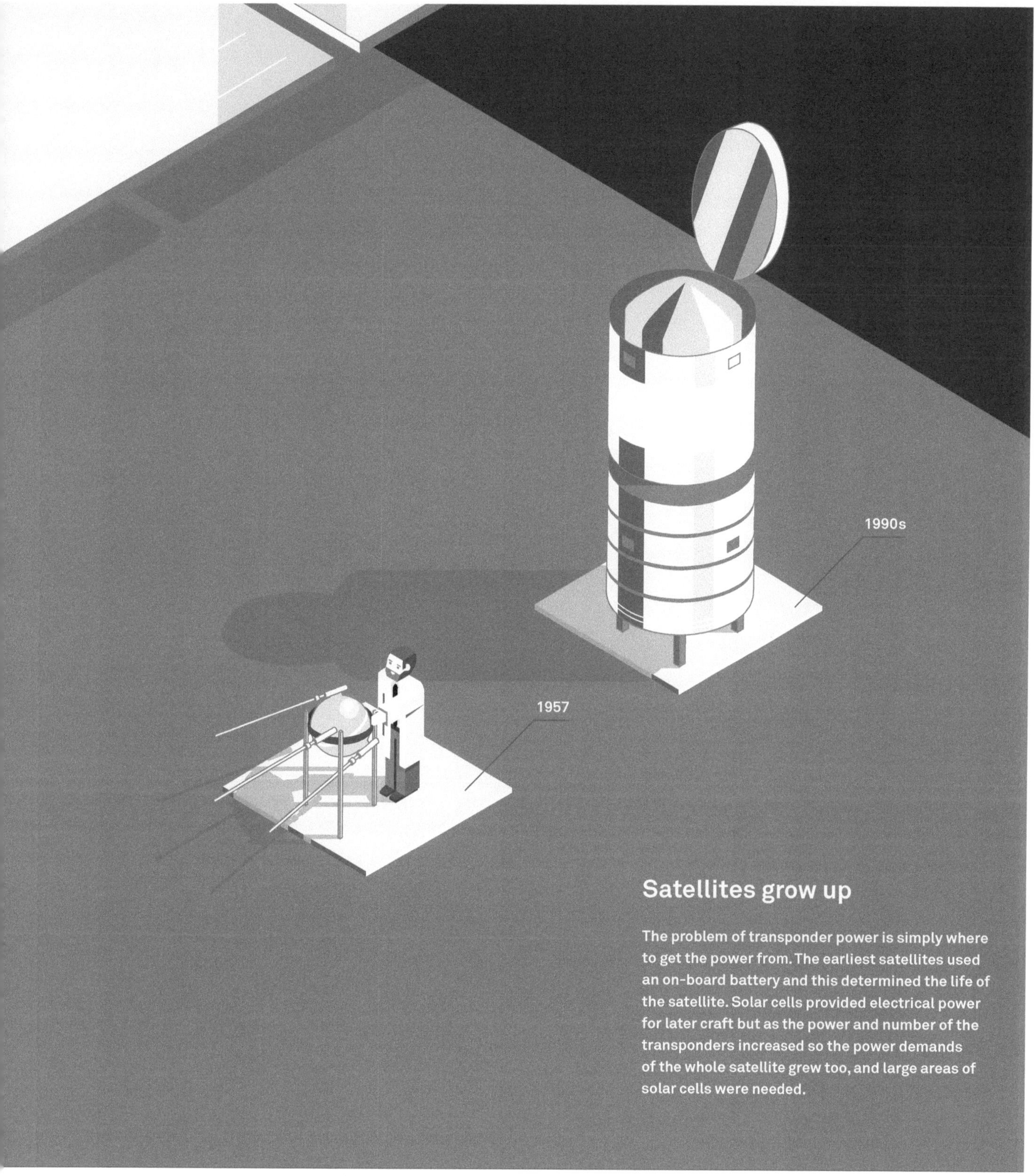

Satellites grow up

The problem of transponder power is simply where to get the power from. The earliest satellites used an on-board battery and this determined the life of the satellite. Solar cells provided electrical power for later craft but as the power and number of the transponders increased so the power demands of the whole satellite grew too, and large areas of solar cells were needed.

02/06/18
SES Americom signs four launch contracts

02/05/26
Mars Odyssey finds signs of ice on Mars

02/06/30
Brazil wins World Cup for a record fifth time

02/07/14
President Jacques Chirac assassination attempt on Bastille Day

13

The Satellite Miracles

13 The Satellite Miracles

Geoff Bains and Chris Forrester

Digital satellite television broadcasts already cover the globe. These broadcasts fulfil two main objectives, and dozens of other objectives. We all recognise the Direct-to-Home broadcasts that beam into homes via a dish or Satellite Master-Antenna System (SMATV) or perhaps are retransmitted via a microwave-based delivery systems.

Then there are the cable systems where the signal is drawn down to a cable head-end that might also be an IPTV-based DSL or similar telephony-delivered signal stream. Finally, there are the rebroadcast terrestrial systems, usually referred to as Digital Terrestrial TV. Very successful examples now exist in almost every European country.

Originally, communication satellites were designed to achieve two-way traffic between far distant communities – the United States and Europe, or Japan and the United States, for example. All the early satellites (from the mid-1960s until the early 1980s) worked to this concept: Telstar, Early Bird/Intelsat 1 are examples.

Few observers envisaged that the size or complexity of the huge Earth Stations then required receiving satellite transmissions could be reduced to the 40-cm or 60-cm dish, able to be casually fixed by relatively unskilled technicians to the side of a house. Another indication of the dramatic change in satellite broadcasting is that the Early Bird satellite weighed just 38 kilograms. The latest Boeing or Loral communication satellites weigh in at almost 6 tonnes.

Today, thanks to these dramatic changes, television now takes by far the largest share of the available satellite capacity. In 1982 there were only about 50 television stations in the whole of Europe, mostly publicly owned. Today there are many thousands, by far the majority being delivered by satellite. An authoritative report from 'Screen

02/09/13
Astra 1A, semi-retired but still working at 5.2 degrees East

02/09/23
First public version of Mozilla Firefox ('Phoenix 0.1') released

02/10/09
Dot-com bubble bursts. Dow Jones falls below 7,200

C. Forrester (ed.), *High Above: The Untold story of Astra, Europe's leading satellite company*,
DOI 10.1007/978-3-642-12009-1_13, © Springer-Verlag Berlin Heidelberg 2011

Satellite is vital, especially in remote places like Mongolia

02/11/25
Astra 1K launched, but rocket fails. Satellite de-orbited in Dec.

02/10/12
Bali bombings. Hundreds die

02/10/23
Chechen rebels take 763 hostages in Moscow theatre

02/12/09
United Airlines files for bankruptcy

Digest' (August 1999) stated that European TV channels have been growing "at over 40 per cent per year since 1995...." That growth continues. 2009 saw a net growth of some 200 new channels over Europe.

Satellite's advantages

Satellite's advantages for the transmission of TV programs are straightforward. A satellite signal can be received on relatively low-cost home receivers – around 100 euros for a simple digital receiver, or for around 200-500 dollars for sophisticated units depending on specification. The ex-works prices of these digital Integrated Receiver Decoders (IRDs) are further falling as production volumes increase and component prices fall. In addition, economies are being achieved by the consolidation of internal components, improved design functionality of chip-sets and new suppliers coming into the market helping reduce the cost of such products.

Satellite can also span thousands of miles within its footprint, delivering signals at a lower cost per household passed than any other method. While broadcasters have to deliver their signals to the orbiting satellite (uplinking) and lease transponder space from satellite operators such as Astra, the remainder of the investment required to view those signals is generally paid by the subscriber, firstly in the purchase of the receiver and dish, and secondly, in subscription fees which are of more direct benefit to the broadcaster.

There's also another payment model, drawn from experience in the cellular telephone business. Receiver prices are increasingly being lowered by broadcasters subsidising, in one form or another, the direct cost of the box. Examples are Canal Plus' lease-payment plan, DirecTV and Echostar's established subsidised box schemes.

Moreover, in the same way that some countries have embraced cellular telephone technology as a means of leapfrogging an inadequate hard wire system, so satellite can provide a near-instant solution enabling countries to completely bypass cable, whatever cable's merits. Such actions can effectively deliver 'free' television to huge areas with little or no infrastructure investment needed by local companies, municipalities or governments.

Satellite can also be received by a Satellite Master Antenna system (SMATV), usually to a group of homes or apartment block. The system receives its signals by satellite and redistributes those signals to individual households either in one building or many adjacent buildings via cable.

Much has been stated of cable's ability to deliver near-limitless data over cable-specific modems. However satellite is beginning to offer very similar capability. Most major satellite operators now offer coordinated receiver equipment for two-way Broadband delivery, supplying all the vast choice of the Internet to just about any home – or business – under the satellite's giant footprint.

2003

03/01/30
Belgium recognizes same-sex marriage

03/02/01
Space Shuttle 'Columbia' disintegrates over Texas. All 7 astronauts die

Dishes are everywhere in Cairo, Egypt.
The famous Giza pyramids are in the far distance.

Satellite can also span thousands of miles within its footprint, delivering signals at a lower cost per household passed than any other method.

03/06/07
SES Americom's AMC-9 launched from Baikonur

03/03/22
US and UK begin 'shock and awe' air strike on Baghdad

03/04/14
Human Genome Project completed

03/06/02
European Space Agency's Mars Express probe launches

Dishes on the Equator look directly overhead

Why is this dish not sloping?

Satellite dishes must point directly to the satellite. Most homes use an offset dish, where the incoming signal bounces off the dish and is focussed onto the Low Noise Block (LNB), the small device that collects and amplifies the signal. This offsetting of the signal allows the whole dish to collect the signal, without the LNB being in the way.

Dishes can also be 'prime focused' where the complete apparatus is looking directly at a satellite.

However, as one gets closer to the Equator then dishes increasingly point straight up into the sky. The satellite orbital arc is directly over the Equator.

Dishes in these equatorial regions are usually (but not always) made of mesh. Mesh is undoubtedly best, especially in the tropical regions, and allows rain to easily drain through.

But there are problems. Dishes that look straight up, whether mesh or solid, make great bird's nests! Storks, and other birds that like lofty nests, love satellite dishes and often make their homes in them.

Europe's 'offset' dishes also make it easy for dish installers to easily set up and adjust for optimum reception. An Astra-equipped home in Germany or France is 'looking' at Astra's 19. 2 degrees East orbital location, and the satellite is positioned over Zaire (the Democratic Republic of the Congo), in Africa and more or less above the town of Ingende. Astra's 28.2 degrees East orbital position is a little further East, almost above Lake Edward on the Zaire/ Uganda border.

03/06/16
Astra orders two satellites from Lockheed-Martin

03/07/03
Astra now in four out of five digital homes

03/07/30
Last-ever 'old style' VW Beetle produced, Mexico

03/08/11
Paris heatwave sees temperatures hit 44°C (112°F)

Astra pioneered a revolutionary technique to extend the capacity available to customers without risking too many transponders at a time. The principle of co-location places two or more satellites in almost the same orbital position.

The Astra Ingredients

Astra started with one orbital position, at 19.2 degrees East, and has grown its presence in space as the popularity of satellite TV in Europe and the demand for Astra capacity has increased. Today there are Astra satellites at five separate orbital positions above the Equator, providing thousands of TV and radio channels to households across Europe. But operating a single satellite at a single location was never Astra's idea of maximising resources.

Of course, the main technique employed to satisfy more customers with each satellite is that used so successfully by Satcom 1 – to provide more transponders, as more transponders mean more TV channels can be carried. The Astra satellites have provided an ever increasing number of transponders on successive craft with the most recent satellites offering around 56. Astra pioneered a revolutionary technique to extend the capacity available to customers without risking too many transponders at a time. The principle of co-location places two or more satellites in almost the same orbital position, close enough together for a dish on the ground to receive signals from both satellites at the same time. In this way the entire frequency spectrum can be made available to TV broadcasters, spread across a huge number of transponders on several satellites. At the 19.2 degrees East position, Astra has employed up to eight satellites and can provide over 100 transponders.

Co-location is used at each of Astra's five key orbital locations to provide many more transponders than could be made available with a single satellite. Crucially, Astra's co-location policy has a second huge benefit to the satellites' customers. Each satellite in orbit not only supplies additional transponders for customers' use but also provides backup capacity in case of a fault, or even an impending fault, with a transponder on any of the satellites at that position. The Astra satellites are designed with flexibility in mind so that one transponder can be instantly switched into use to replace another, seamlessly and without the viewers on the ground even aware of the change.

In the highly competitive world of television broadcasting, even a temporary failure of delivery can mean the permanent loss of your audience, and so Astra's policy of providing in-orbit backup capacity has endeared it to broadcasters looking for a reliable means to reach – and keep – the huge audiences they desire.

These huge audiences were first growing in the 1970s, in the USA, where satellites not only delivered TV programming to cable networks and local affiliate stations, but also to the first home satellite dishes. It was the lack of access to cable networks by isolated viewers in rural America that prompted the development of home satellite dishes, initially to tap into the cable feeds serving the cities.

These first domestic satellite TV systems were for C-band reception – that is, signals in the range 3.7-4.2GHz – as that was the frequency band used for TV distribution. C-band required dishes typically 3.5 metres in diameter but so long as home reception was a semi-illicit spin-off of cable TV distribution, convenience of reception was not regarded as important. However, as more home dish owners sprang up and the broadcasters saw there was a new market for their

Dishes can even be comfortable: Hidden antenna from TechniSat

03/09/09
Astra confirms launch contract for Astra 1KR

03/09/12
Astra showcases HDTV at IBC technology show

03/09/10
Anna Lindh, Swedish Minister, fatally stabbed

03/09/12
Johnny Cash, musician, dies

A typical C-band dish

programming, consideration was given to making home reception easier and cheaper. The first result was an increase in transponder transmission power to reduce the size of dish needed.

Later, in the 1980s, as the technology and desire for broadcasting TV directly to people's homes spread to Europe, the satellite designers turned from C-band to Ku-band (10.50-12.75GHz), where there was more 'room' for a greater number of channels, particularly as, at that time, the European countries had plans for their own national TV satellites and services.

Television satellites have grown steadily more powerful as transmitter technology has improved and the demand for smaller and more convenient dishes emerged. The first Astra satellite provided 16 Ku-band transponders of 45W, which were receivable across most of Central Europe with a dish 60 cm in diameter (nearly half the size of dishes needed for other satellites at the time), with an 80-cm dish required towards the edges of the footprint.

The latest communications satellites use 150 W transponders, providing reception on 45-cm dishes over most of Europe and on 80-cm dishes far into the Middle East, Russia, Scandinavia and North Africa. The relentless pursuit of technical improvements has enabled far cheaper and smaller dishes to be used, which in turn has popularised satellite TV and extended its potential market to an extent barely dreamt of just a few decades ago.

03/10/23
Astra and Euro1080 announce European HDTV channel

03/10/07
Arnold Schwarzenegger elected Governor of California

03/10/24
Last Concorde commercial flights

In 50 years, TV satellites have come from experimental military spin-offs to a vast, global industry, providing tens of millions of homes with a choice and quality of television never seen before.

Dishes in Neukölln, Berlin

03/12/15
SES increases stake in NSAB to 75%

03/12/13
Saddam Hussein captured by US troops

03/12/26
Massive earthquake in Iran. 40,000 killed

In addition to new orbital positions, communications satellites now increase the capacity available to customers by 're-using' their frequencies from a single position, even from a single satellite. A secondary transmission beam is directed to a region outside the main target reception area for the satellite. By tightly focusing the transmissions on each beam, they are isolated and the satellite can re-use the frequencies of one beam on the other, for different services, without interference to receivers in either area.

At the Astra 5 degrees East position, for example, the Sirius 4/Astra 4A satellite (launched November 18, 2007) provides different direct-to-home TV and radio broadcasts to both Europe and sub-Saharan Africa on the same frequencies, using two directed beams. At the Astra 28.2 degrees East position, the Astra 2B satellite (launched September 14, 2000) provides a main beam to the UK and Republic of Ireland for direct-to-home digital television and a second, steerable spot beam that uses the same frequencies to provide telecommunication services to West Africa via SES Astra's sister company SES World Skies.

Such frequency re-use is likely to become a crucial means to expand satellite capacity worldwide as the demand for satellite communications increases, and it complements Astra's existing co-location policy well. Communications satellites are also expanding available capacity by using new frequency bands for satellite transmissions. For example, the S-band (2.0-2.2GHz) is being developed by the Astra-Eutelsat joint venture Solaris Mobile to provide TV, radio and two-way communication to a variety of handheld and vehicle-mounted mobile devices. A growing number of satellites is also provided with Ka-band (18-31GHz) transponders which are currently mainly used in order to further develop satellite broadband services.

So, like the satellites in orbit, satellite television has to an extent, come full circle; satellites are still providing point-to-multipoint relays of TV signals more efficiently and cheaply than any terrestrial system can manage, but satellites are also finding new ways to help people reach one another and push the boundaries of communications ever forward.

In 50 years, TV satellites have come from experimental, military spin-offs to a vast global industry, providing tens of millions of homes with a choice and quality of television never seen before.

Like the satellites in orbit, satellite television has, to an extent, come full circle; satellites are still providing point-to-multipoint relays of TV signals more efficiently and cheaply than any terrestrial system can manage.

2004

04/01/08
Queen Elizabeth II launches RMS Queen Mary 2 liner

04/02/02
Americom-10 launched from Cape Canaveral

04/02/04
Facebook founded at Cambridge, Massachussetts

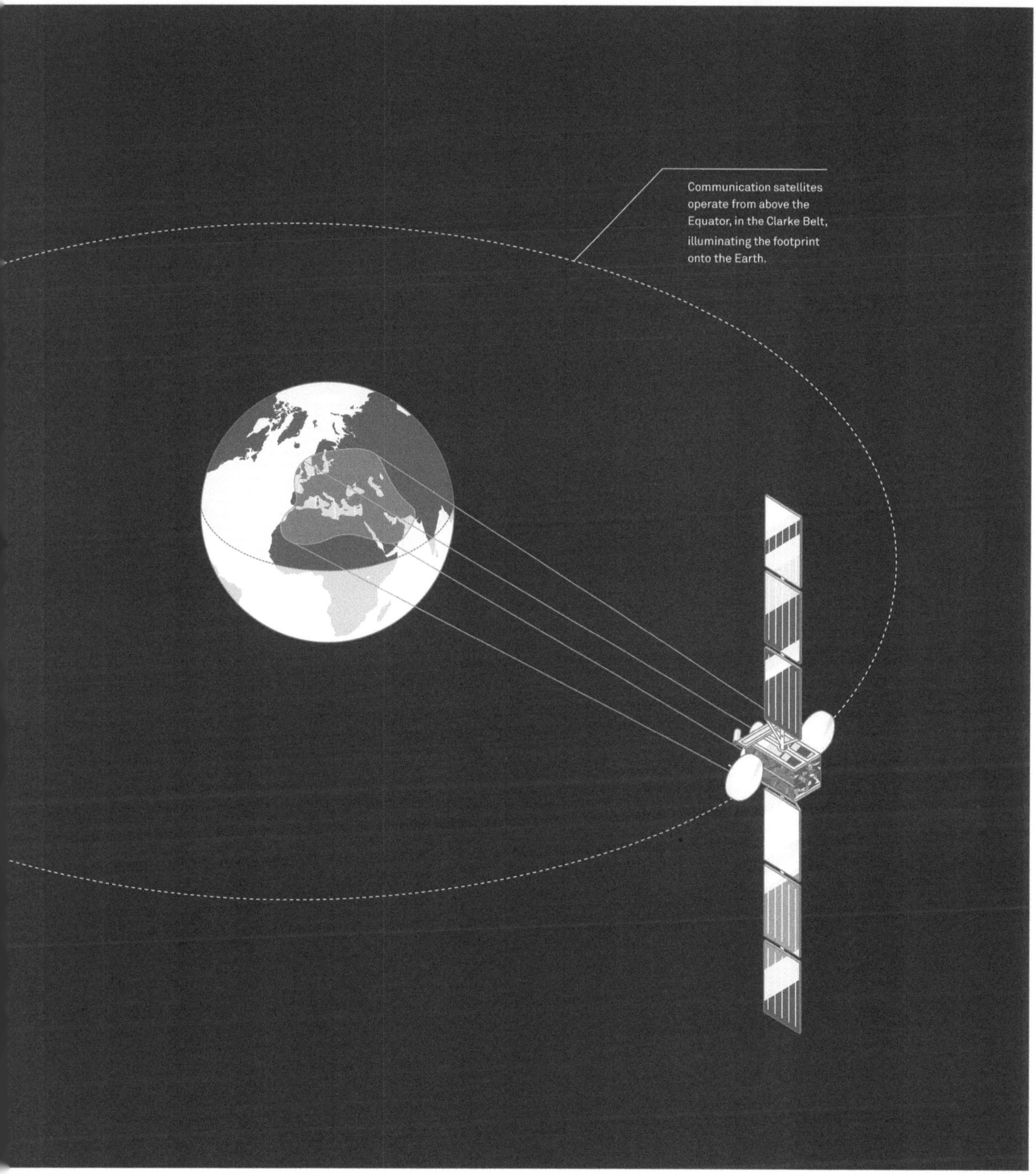

04/03/14
Vladimir Putin easily wins a second term

04/05/01
EU expands with ten new members

04/05/11
Astra extends reach to over 94 million homes

04/05/20
Americom-11 launched from Cape Canaveral

04/06/05
Ronald Reagan, 40th President, dies

14

The TV Revolution: Beyond the Shannon Limit

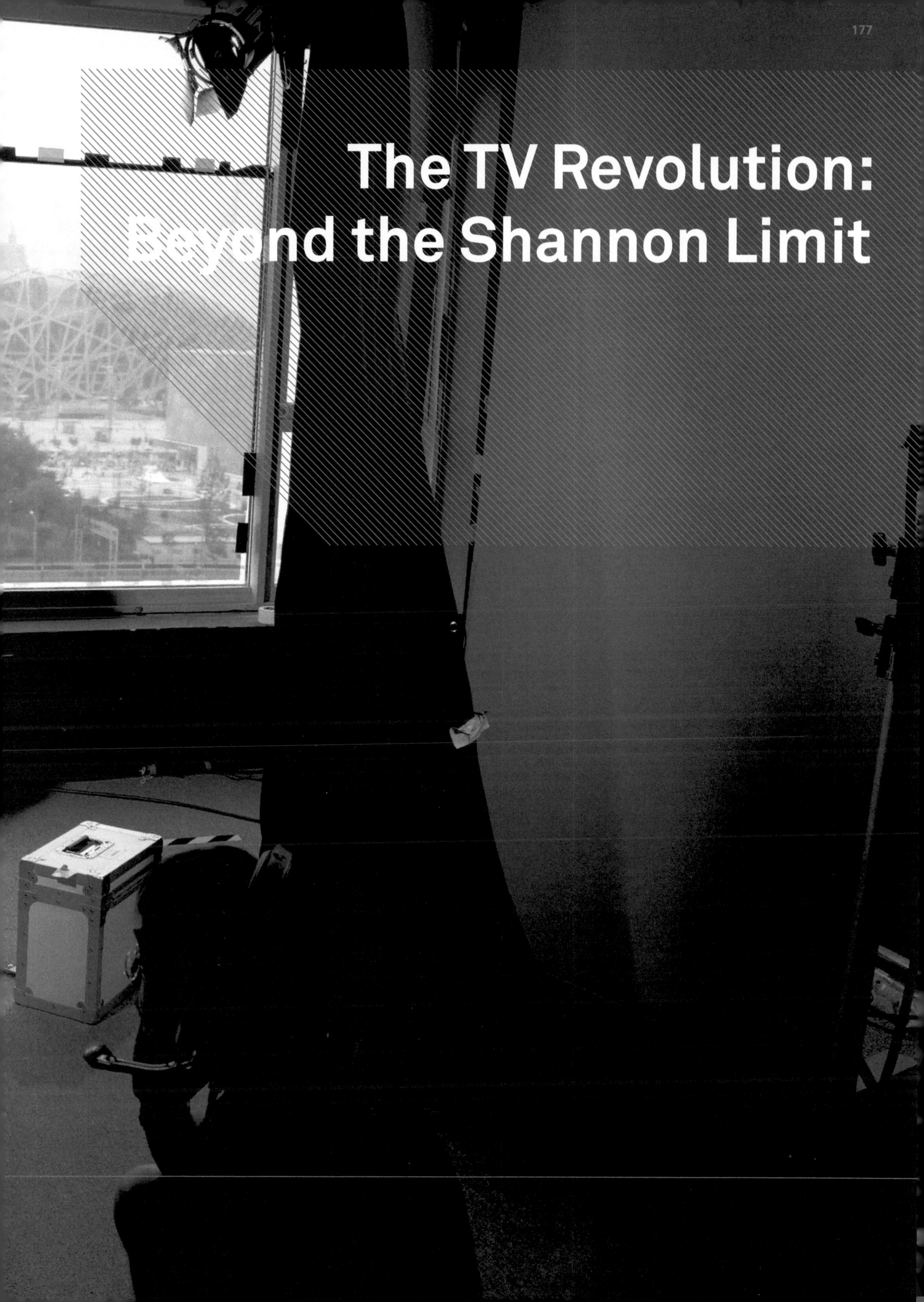

14

The TV Revolution: Beyond the Shannon Limit

Julian Clover and Chris Forrester

Ask any of the video compression engineers what their own Research & Development departments are doing in terms of 'next generation' ideas and they tend to immediately fall silent. It isn't that they are reluctant to speak, but they are cautious about giving too much away to potential rivals. However, the key compression players like Ericsson/Tandberg, Cisco, Harmonic/Scopus and Snell are all working hard on their 'super secret' ideas that include taking a fresh look at compression technology. To succeed they have to take us beyond the theoretical extremity known as the Shannon Limit.

Indeed, the future development of broadcasting methods such as 3D-TV and so-called Super or Ultra High-Definition all depend on advanced compression methods emerging. The progress in compression technology has already indefinitely multiplied the number of channels distributed over satellite – from just 16 analogue channels in 1989, to an estimated total today of around 7,200 digital TV channels and services on European satellites at the end of 2009, and growing by about 200-250 a year. From quantity to quality: this progress will also dramatically increase the data rate – and therefore the quality of the programs delivered. And the development does not stop here.

DVB-S: satellite's standard of choice

If it was the DVB-S standard that gave satellite broadcasting the impetus in the launch of digital broadcasting, then DVB-S2 took it into the 21st Century. The Geneva-based Digital Video Broadcasting Project (DVB) encompasses some 280 broadcasters, manufacturers, network operators, software developers and regulatory bodies across 35 countries.

Broadly speaking, the DVB portfolio of standards has been the driving force behind the growth of multi-platform television, both in

04/07/28
Astra buys 75.2% stake in Digital Playout Centre (DPC)

04/09/01
Astra starts HDTV demo channel

04/10/04
Pro7/Sat1 broadcasts first HDTV/SD simulcast

04/07/15
Lance Armstrong wins unprecedented sixth Tour de France

04/08/13
Athens Olympics open

04/09/22
TV series 'Lost' airs pilot episode

C. Forrester (ed.), *High Above: The Untold story of Astra, Europe's leading satellite company*, DOI 10.1007/978-3-642-12009-1_14, © Springer-Verlag Berlin Heidelberg 2011

A technician makes final adjustments prior to the giant IFA trade show in Berlin

From quantity to quality: this progress will also dramatically increase the data rate – and therefore the quality of the programs delivered.

Europe and the majority of countries outside of North America and Japan. Even in the United States, which opted for its own ATSC digital broadcasting standard, the DVB-developed technology is used for satellite transmission as well as contribution feeds for news-gathering, sports transmissions and the like.

Not all the DVB's work has been a success; the MHP middleware has largely failed to make an impact, as patent holders piled on the costs, though it has emerged in the United States as the interactive standard 'tru2way' for cable and within the Blu-ray software stack. There are also doubts about the DVB's role in IPTV standards, a result in part of the involvement of the telcos, which have chosen to plough their own technological furrow.

The DVB was formed in 1993, emerging out of the European Launching Group, which had been established two years earlier. DVB-S

04/10/04
SES Americom's AMC-15 launched from Baikonur

04/10/17
SES Americom's AMC-16 launches from Cape Canaveral

04/10/19
EU Constitution signed

04/11/11
Yasser Arafat dies in France

04/12/26
Sumatra tsunami kills 187,000

The Shannon Limit

The Shannon Limit (or more accurately, Shannon's fundamental theorem of information technology) describes a concept first presented by the brilliant US electronics engineer and mathematician Claude Shannon in 1948, and explains where the compression of data within a channel will achieve a natural limit. Beyond this point, the amount of noise – or data errors – will overwhelm the original signal. The mathematics behind the theory were amplified by Amiel Feinstein in 1954. TV engineers are keen to exploit the compression possibilities below the Shannon Limit, where it is perfectly possible to encode and transmit data without error at all.

came about in 1993 and combined with MPEG-2 compression technologies, became the satellite standard of choice.

Fast forward to 2003 and with the majority of Europe's satellite channels transmitting digitally – even with an analogue hangover that remains today – attention turned to greater spectrum efficiencies that would be needed if Europe was to follow the United States into high-definition broadcasting. Indeed, the transmission technology originally grew out of the idea to deliver HDTV in the United States over its 26 MHz – not 36 MHz – transponders.

The resultant DVB-S2 technology is backwards compatible with the previous DVB-S – even if the early receivers required two chipsets that clunked as their tuners moved from standard to high-definition channels – allowing for a data capacity 30% greater than with DVB-S on a receiving dish of the same size. It was ratified by the European Television Standards Insitute (ETSI) as EN

May 10, 1952: Dr. Claude E. Shannon, a mathematician at the Bell Telephone Laboratories with an electronic mouse which has a 'super' memory.

DVB-S3 is not completely out of the question, but it seems unlikely it will be on the agenda for some time to come.

302307 in March 2005. The work coincided with the introduction of H.264 (MPEG-4 AVC) video codecs.

When DVB-S2 was on the drawing board, one of the requirements was that it fitted into the same area on the silicon as the technology it was replacing, in 2003 this meant about 14 mm squared. The combination of the latest modulation and coding techniques took DVB-S2 close to the Shannon Limit. In other words a DVB-S3 is not completely out of the question, but it seems unlikely it will be on the agenda for some time to come.

The first end-to-end DVB-S2 appeared at IBC 2005 when Tandberg Television, fresh from performance tests with the European Broadcasting Union, unveiled its TT1260 SD MPEG-2 receiver with an integrated DVB-S2 demodulator and proved that a savings of 30% could be made within a given bandwidth.

While recent advances in transmission may leave little room for improvement, engineers at the BBC's Kingswood Warren research and development facility have turned their attention to video compression. 'Dirac' was originally developed to deal with high-resolution images and higher bit rates, but has subsequently been proven to be suitable for applications running at over 1 Gbit/sec and below 100 kbit/sec. This puts it on a par with H.264/MPEG-4 AVC and VC-1.

Dirac, in its basic form, is open source software and consequently not hindered by the royalty issues that held back the MHP application programming interface. Its uses are wide and suitable for the Internet delivery of video clips and onto high-definition and even digital cinema. Dirac technology was also originally selected to play a part in Japan's development of its Super Hi-Vision and Ultra HDTV technologies. Dirac may also be a key to the delivery of the on-demand content that is expected to play an increasing role in video consumption over the next decade

TV screens at the IFA Berlin international trade fair

05/02/15
Internet site YouTube debuts

05/04/02
Pope John Paul II dies

05/04/08
Astra issues contract for Astra 1M

05/04/13
Astra increases stake in ND-Satcom to 25.1%

Digital images are good – but can also create problems: digital blur on a TV screen

as the BBC continues to champion offerings like its iPlayer download service. What started as a web-based catch-up TV service seems increasingly likely to sit side by side with the distribution of linear channels over satellite, cable and terrestrial platforms.
A professional version of Dirac has also been developed, enabling the transportation of HD content at 50 frames per second, so double that of what has previously been achieved and adding a more 'fluid' motion to the images. Dirac Pro is said to give little or no loss of quality and is particularly suitable for the delivery of sports content. It could be in use for the 2012 Olympic Games in London, for example.

It is evident that in a hybrid constellation, satellite can again fully leverage its advantages as the most powerful broadcast infrastructure.

05/04/26
Astra reach tops 102 million homes

05/05/05
UK Labour Party re-elected

05/06/28
Trafalgar Day 200 anniversary celebrated by Queen Elizabeth

It was demonstrated at IBC in 2006 with the improved Dirac Pro 270 appearing at the same show 12 months later. It has also been seen by the Hollywood Post Alliance and is being considered for standardisation by the Society of Motion Picture and Television Engineers (SMPTE).

After DVB-S2 and Dirac Pro, it's 3D

In April 2009, SMPTE said it would undertake the generation of specifications for a 3D Home Master that would create a single feed into the home and carry content from mobile, DVD, streaming, terrestrial and cable as well as satellite devices. The key requirement is for a video resolution of 1920×1080 and native frame rates up to and including 60 (progressive) frames per eye view. The importance of backward compatibility to existing 3D content has been underlined by the organisation.

SMPTE plans to work with other relevant standards development organisations, as well as industry consortia and forums. In Europe alone, a variety of organisations are looking to exert their influence on the emerging 3D market. These include the DVB, EBU and Digital TV Group.

In a report published in June 2009, the DTG members said they feared that de facto, non-open standards would emerge from technology decisions being made by the first providers, most likely pay-TV operators, which would impact later free-to-air broadcasts.

Already there is friction between Ofcom and Sky over the development of 3D technologies. The regulator favours a single standard, but no sooner has the technology left the drawing board than Sky in the UK has taken up every opportunity to show its technology to technical and investor groups. In addition to demonstrations by companies with their own proprietary technology, the French IPTV operator Orange is also a major proponent, capturing the French Open tennis in the format for two consecutive years.

3D television images being demonstrated by NHK in Tokyo

05/07/05
SES elected to CAC 'Next20' and 'IT20' indices

05/07/07
Bus and tube atrocities in London kill 56

05/07/21
Astra awards Astra 1M contract to Astrium

05/07/20
James 'Mr. Scott' Doohan, Star Trek star, dies

There's also considerable work going on with the so-called HbbTV concept of hybrid broadcast broadband TV. HbbTV is a major new pan-European initiative aimed at harmonising the broadcast and broadband delivery of entertainment to the end consumer through connected TVs and set-top boxes. The HbbTV specification was developed by industry leaders to effectively manage the rapidly increasing amount of available content targeted at today's end consumer. It is based on elements of existing standards and web technologies including OIPF (Open IPTV Forum), CEA, DVB and W3C.

HbbTV products and services will provide the consumer with a seamless entertainment experience with the combined richness of broadcast and broadband. This entertainment experience will be delivered with the simplicity of one remote control, on one screen and with the ease of use of television that we are used to. Through the adoption of HbbTV, consumers will be able to access new services from entertainment providers such as broadcasters, online providers and CE manufactures – including catch-up TV, video on demand (VoD), interactive advertising, personalisation, voting, games and social networking as well as programme-related services such as digital text and EPGs. It is evident that in a hybrid constellation, satellite can fully leverage its advantages as the most powerful broadcast infrastructure.

BSkyB has gained the chairmanship of the potentially influential DVB working party on 3D broadcasting. According to the European Broadcasting Union there are currently at least six different approaches emerging for the broadcast of 3D-TV. This in itself brings challenges as technical minds look to find a format that is capable of working with all of the available display types. Generally speaking, the choice is between active and polarised stereoscopic systems and single-view auto-telescopic that adds depth to 2D broadcasts. To this, different screen sizes have to be taken into account because the view from the front row of the cinema may be very different to that enjoyed at home.

Introducing 3D asks the question as to where all the bandwidth is going to come from, as consumers demand a more personalised experience. The answer, so far, has been in the hybrid set-top receiver which combines satellite, and possibly even a co-axial connection, with IPTV delivery forms.

Introducing 3D asks the question as to where all the bandwidth is going to come from, as consumers demand a more personalised experience.

A successful implementation has been through the Canal+ product Le Cube. The next generation receiver combines a personal video recorder that pulls its linear content from the French pay-TV provider's Astra-based platform with an IPTV service. Canal was swift to implement the technology to compete with the Orange IPTV service that has quickly moved to 1.7 million subscribers, as of 2009.

With these hybrid applications, fascinating new forms of broadcasting are appearing on the horizon. 'Broadcasting' is no longer a single method of distribution from satellite (or broadcast tower) to the viewer but, like the movies, is undergoing a revolution in the number of distribution methods. Viewers and consumers today expect to see their favourite content accessible just about everywhere. Multiple ways of delivery on multiple devices, based on invisible transmission standards that are as open as possible: The so-called 'Martini' moment ("Anytime, anyplace, anywhere") might never have been true for the drink, no matter how popular it is. The reality is that programming has truly reached the 'Martini' moment in terms of accessibility. With its reach and its technical capabilities, satellite will be key for these developments.

05/08/29
Hurricane Katrina hits US. 1,800 die

05/09/09
BSkyB takes four extra transponders from Astra

05/09/12
eBay buys Skype

05/12/14
SES agrees to acquire New Skies Satellite for $1.1 billion

05/10/01
Bali bomb outrage. 26 killed

05/11/30
French surgeons carry out face transplant

05/12/31
Extra second added at midnight

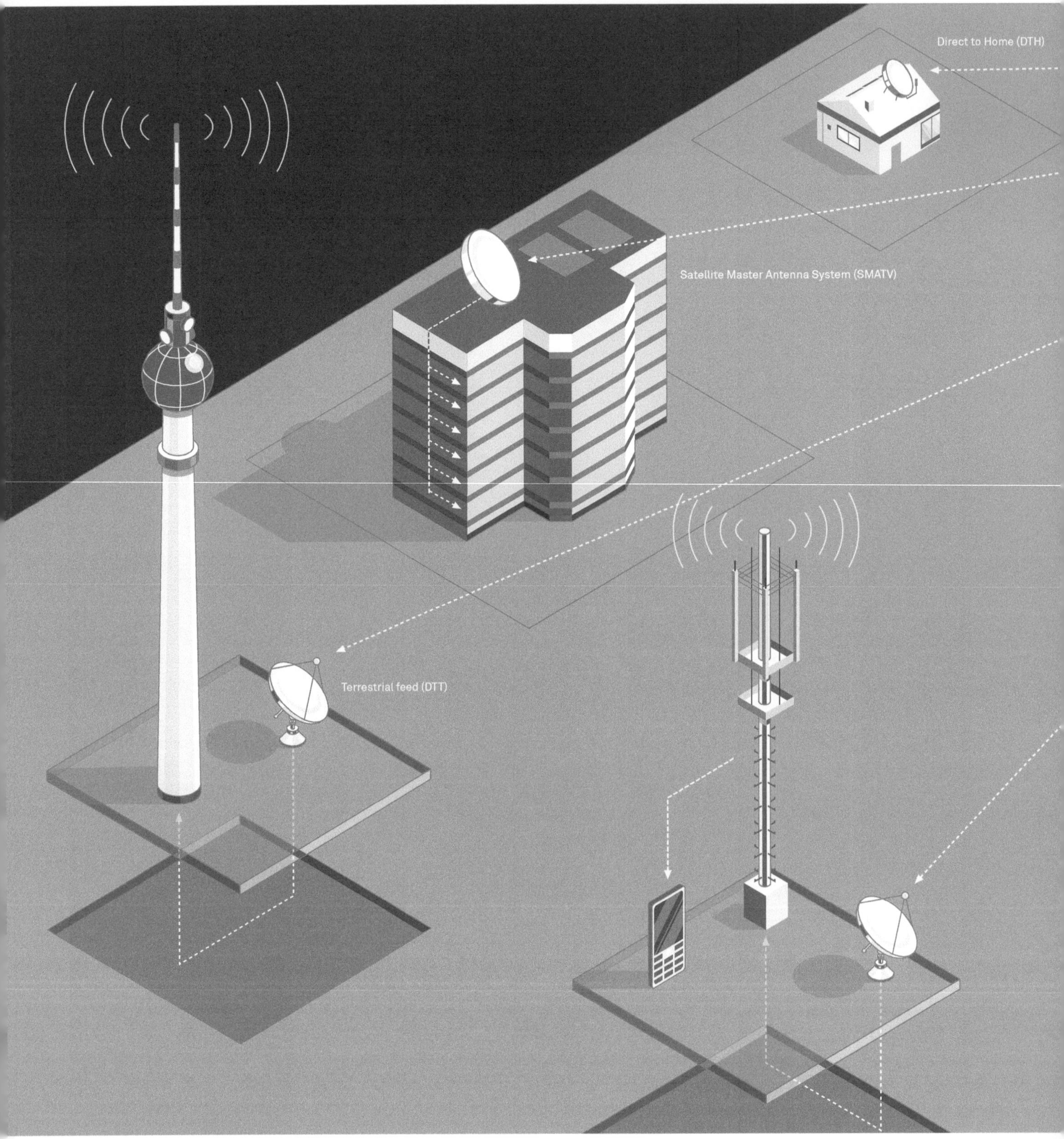

2006

06/03/02
SES Americom's AMC-23 operational

06/03/30
SES completes purchase of New Skies Satellite

06/01/27
250th birthday of Wolfgang Amadeus Mozart

06/02/10
Winter Olympics in Turin, Italy

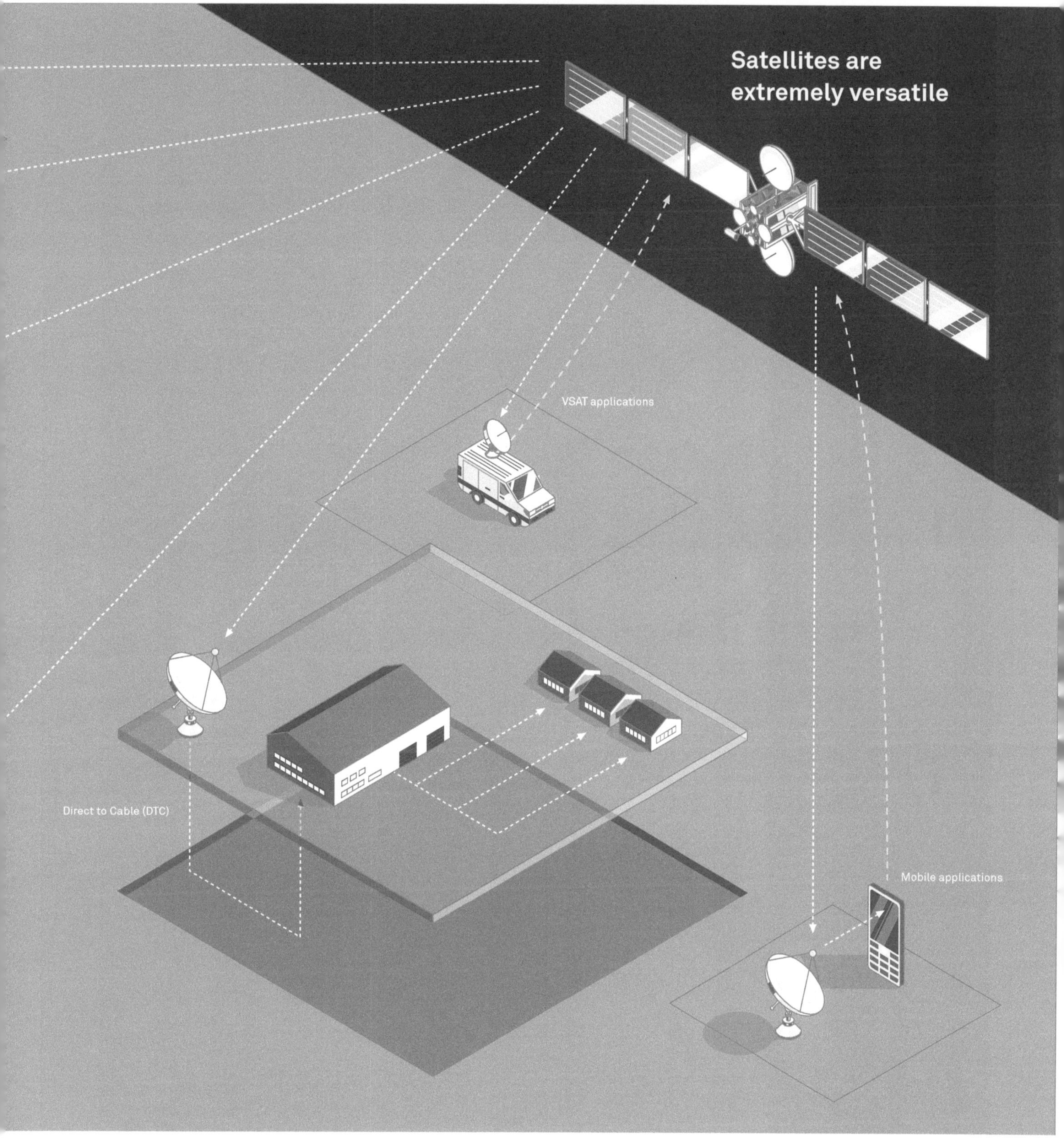

06/04/06
Canal+ takes extra HDTV capacity

06/04/21
Astra 1KR launched from Cape Canaveral

06/05/03
SES Americom orders AMC-21 from Alcatel-Alenia

06/03/16
Blu-ray format released in the US

06/06/18
Kazakh space satellite KazSat is launched

15

The Future for the Box

15 The Future for the Box

Julian Clover and Chris Forrester

The humble set-top box has come a long way over the past 25 years. Back then it was still possible to purchase a satellite receiver, complete with a clunky rotary dial for tuning, but alas no remote control. This all changed with the launch of Astra and, in the UK market, the appearance of the first Amstrad receiver as part of the agreement between Rupert Murdoch's Sky Television and (now Lord) Alan Sugar.

There are plenty of TV sets now available which happily provide reception of satellite, cable, digital terrestrial or IP-based transmissions. Indeed, some can comfortably handle multiple inputs from different sources. This element alone might be the next revolution for home viewers.

Through the 1990s, receivers in most markets took on much the same form, with LED displays, RF connectors to feed the signal through to the UHF bands used by the broadcast channels and SCART/Peritel sockets to take advantage of the direct connection required by the European Commission on displays with larger screen sizes.

The key analogue innovation came after Sky switched on the encryption for its movie and sports channels. When Sky Sports was launched after the merger with British Satellite Broadcasting, the introductory offer included a Thomson-manufactured standalone VideoCrypt decoder. It was getting crowded underneath the television set!

Manufacturers were free to develop receivers as they saw fit, gradually improving the on-screen graphics and general functionality. But when digital services started to emerge towards the end of the 1990s, the platform operators began to take control, even so far as to specify a set-top box design and in the case of Canal+, further insisting that all receivers that carried its brand would have exactly the same core chassis. It would be ten years before Sky imposed

06/06/16
Astra 1KR operational

06/06/30
Germany approves Astra purchase of ND-Satcom

06/10/30
Astra and Eutelsat cooperate on Solaris Mobile

06/07/12
Israeli troops invade Lebanon

06/09/19
State of emergency in Thailand

C. Forrester (ed.), *High Above: The Untold story of Astra, Europe's leading satellite company*,
DOI 10.1007/978-3-642-12009-1_15,

the same design rules on its manufacturers, but it already had plans to provide a uniform feel to its service. While OpenTV and NDS worked on the look and feel of the electronic programme guide, Universal Electronics produced a remote control that would allow Sky to give viewers a simple explanation as to how everything worked, knowing that when customer services told subscribers to press a particular button, they could have a reasonable confidence that everybody would press the same one. The remote was replicated by News Corp affiliates across the world from Foxtel in Australia through to Sky Italia.

When customer services told subscribers to press a particular button, they could have a reasonable confidence that everybody would press the same one.

The Sky Digibox added new card-reading slots and connections to back and front. Two smart card slots offered room for both the smart card, giving entitlement to the various channels to which the viewer had subscribed, and a slot for a so-called interactive card. Early on, Canal+ devised a scheme whereby French viewers – who were used to a chip being present on their credit and debit cards – could obtain credits to use for pay-per-view purchases. Instead, Sky made it an obligation that in order to benefit from its initial digital set-top box subsidies – taking the cost first to 200 pounds and later free – that subscribers connect their receiver to a telephone line for the first 12 months of their contract.

It wasn't until 2005 that Sky made use of the interactive card slot, but the launch of the Interactive Credit Card with Barclaycard was even then very short-lived, even the lure of accruing SkyPoints to use against subscription purchases did not prove a strong enough incentive.

There were other elements that began to appear within receivers: High-Definition services brought connectivity through HDMI connections, and the hybrid receivers began to appear. Canal Digital used hybrid receivers in the Nordic market to ensure its subscribers could have seamless access to the free-to-air channels available over DTT platforms where a direct agreement was not in place.

In November 2008, Canal+ introduced a new generation of hybrid HD satellite receivers. Called Le Cube, the receiver brought in broadband access to enable subscribers to have access to on demand content, it was also designed to combat the newly launched Orange DTH platform.

06/11/20
SES orders NSS-9 from Orbital Sciences

06/11/21
Astrium contracted to build Astra 3B

06/12/09
SES Americom's AMC-18 launched from Kourou

06/11/19
The Wii goes on sale in US

06/12/30
Saddam Hussein executed in Baghdad

New interactive experience:
Le Cube from Canal+

2007

07/01/31
NSS-8 lost as rocket explodes on launch

07/01/01
Bulgaria and Romania join the European Union

07/02/20
Astra/Sirius carrying over 2,000 channels

07/03/31
Sydney, Australia turns off its lights for one hour

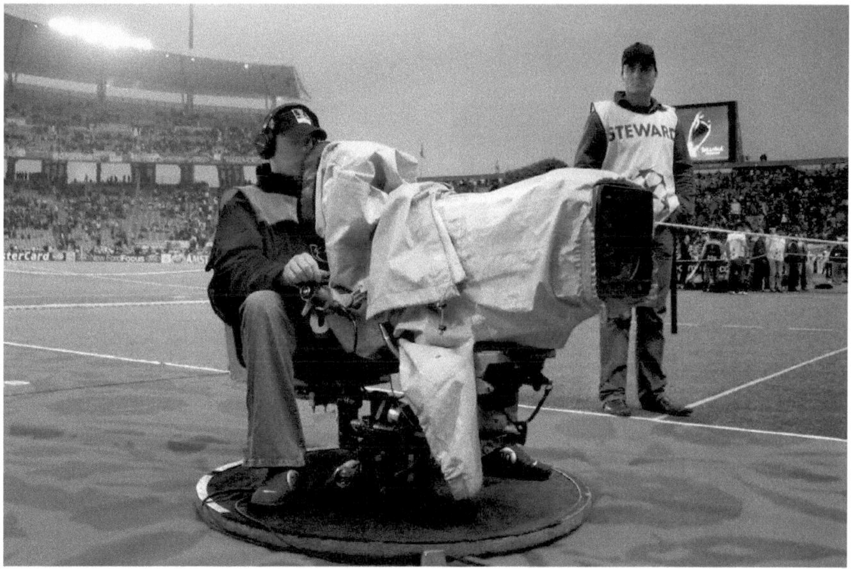

Supplies of Le Cube began shipping on November 4, 2009 and represent Canal's first HD Personal Video Recorder (PVR). Le Cube's hard disk has been optimised by Pace for low audibility, the whirring hard drive noise being a regular criticism of PVR

An IP connection allows progressive downloads of content with programming buffered onto a hard-disk drive.

devices. Its 320GB hard drive allows 100 hours of HD recording. In addition to an Ethernet port, this product supports USB and HDMI and has Dolby Digital audio outputs. The IP connection allows progressive download of Video-on-Demand (VOD) whereby content is buffered onto the hard drive.

Le Cube cost viewers an additional 10 euros a month (the total rental price was initially 18 euros, compared to 8 euros for the current DTH receiver). With the Ethernet connection on the box, Canal+ also offered interactivity, VOD services and new services such as Restart-TV and a recommendation service for viewers.

A further variation to the hybrid box, again involving broadband connectivity, was introduced by Viasat in March 2009. Like Le Cube, Viasat used technology developed by NDS. Subscribers are able to download a movie or series through their broadband Internet connection for immediate viewing or to store for later on. NDS says its progressive download technology delivers uninterrupted viewing by removing the buffering problems associated with slow broadband download speeds.

The Viasat box uses available bandwidth to download content to the receiver hard drive. Once enough content has been cached, viewing can commence, and the remaining content downloaded in the background. The Progressive Download also reduces the need for operators to deploy the high specification VOD servers normally associated with HD content delivery.

The importance of keeping control over its technology was demonstrated by Sky in July 2007, through the purchase of Sir Alan Sugar's Amstrad in a deal worth 125 million pounds. Although the broadcaster continued to use its other two favoured suppliers, Thomson and Pace, the Amstrad acquisition has enabled Sky to control both the design and the cost of its satellite receivers. Amstrad supplies some 30% of the set-top boxes bought by BSkyB including the standard Digibox and Sky+ PVR.

07/03/07
Astra reach now more than 109 million homes

07/05/05
Astra 1L launched from Kourou

07/05/10
SES orders NSS-12 from Space Systems/Loral

07/04/16
Virginia Tech massacre. 32 killed

07/05/20
Clashes in Tripoli, Lebanon spark conflict

Ever larger screens can handle multiple information sources

Sky has also broadened its relationships with chip-set manufacturers and hard-disk providers and connected the division through to the Sky Research & Development team. Such relationships have made reductions to the price of the Sky HD receiver possible.

Current thinking is for a series of connected set-top boxes, perhaps with equal functionality, or maybe a series of slave receivers working from a mother receiver.

However, it is not just pay-TV broadcasters that are eyeing up Internet connectivity. Key to UK free-to-view Freesat's offering is an Ethernet port that will enable connection to broadband services. The appearance of these hybrid Connected TVs, delivering widget-style information without the need for a set-top box, is a potential game changer for the multichannel TV sector, particularly when the concept is opened up to the possibility of micropayments for services that include video.

The pay-TV operators have not stood still; NDS and others are working on concepts that fuse content from the Internet with the broadcast stream. NDS's Infinite TV, for example, is capable of running on a number of devices including PCs, hybrid set-top boxes, integrated digital TVs (IDTVs) and portable media players. Users can register their details and devices online and then view content. Advertising can be tailored to the content viewed, with advertisements appearing when the programme is paused – a functionality that must surely find its way into standard PVRs.

Although multi-platform, NDS has been playing up the potential relationship between broadcasters and consumer electronics manufacturers, at the same time being careful not to promote a low end competitor to its core pay TV constituency.

Sky intends to pick up on shared planner technology developed by NDS to allow a subscriber to be able to view the content available on a central DVR from any set-top box available on the network. A nod perhaps

07/06/19
SES contracts for ten launches with Arianespace

07/07/11
Astra 1L operational

07/07/21
'Harry Potter and the Deathly Hallows' released

07/08/01
Scouting celebrates its 100th birthday

towards the touted multiple tuner device that has been on the Sky drawing board for the last three years.

It also spells a slight shift away from the concept of having a server hidden away under the stairs that feeds content throughout the house. Instead, current thinking is for a series of connected set-top boxes, perhaps with equal functionality, or maybe a series of slave receivers working from a mother receiver that is most likely connected to the main TV display.

Home networking can be expected to emerge over the coming years, taking advantage of emerging standards such as MoCA (Multimedia over Coax Alliance) and the Digital Living Network Alliance (DLNA), the latter with 245 members already committed to the project.

In the United States, Motorola's MoCA enabled set-top boxes allow recorded DVR content to be viewed on a subscriber's home network. Typically, three HD programmes could be played over a network while another two are recorded. The technology is being supported by the interactive standard tru2way supported by the US cable industry under the auspices of CableLabs.

In Germany, the new Astra platform for HD, HD+, offers broadcasters the largest possible flexibility.

In Europe, the emergence of CI Plus out of the DVB Common Interface module has given a further boost to Integrated Digital Televisions (IDTVs) and the inter-operability of set-top boxes. Rather than just allow a different conditional access system to be used, CI Plus adds greater functionality including VOD navigation, rich interactive ads and Web video. Already it has been used by Canal+ to add HD pay channels to the hitherto free-to-air package known as TNT-Sat. The French broadcaster has even created its own label, Canal Ready, to flag the functionality to French consumers.

In Germany, the new Astra platform for HD, HD+, is supporting CI Plus in order to offer broadcasters the largest possible flexibility as well as a state-of-the-art protection of the broadcasters' signal – against piracy, illegal copies and unauthorized use.

The opening up of the devices marks an interesting turn of events. Previously, the TNTSat receivers were solely intended for reception of the terrestrial digital channels via satellite using Viaccess encryption. By deploying the CI Plus standard, they will also open up to other encryption systems, including that of Canal+ and CanalSat.

There is, however, one element that most set-top box manufacturers would happily prefer *not* to talk about. Increased access to broadband services via Ethernet or WiFi links within the actual TV set means that the TV set itself is now able to deliver programming to the viewer. In fact, what is to stop a major vertically integrated Hollywood studio like Sony, with massive investments in movies, music and games, as well as products like Blu-ray and a giant electronics division that's already making high-end TV sets, from moving into programme distribution?

The argument is that an organisation as powerful as Sony (but it could also be Panasonic, or Samsung, or LG) will add a new revenue stream to its current portfolio of interests, and sell 'subscriptions' to the home for access to its programming. Time will tell, but there's a major question as to which company, these days, controls the home's entertainment mix. Is it the likes of Canal+ or BSkyB, or a telephone company, or the likes of Sony, Panasonic or a telephony operator? The next 25 years promise to be just as exciting as the past 25.

07/09/05
SES wins 'Sat-operator of the Year' for second year running

07/10/23
Astra adds capacity to 23.5 degrees East

07/09/01
Finland switches off analogue TV

07/09/06
Luciano Pavarotti, opera star, dies

16

How on Earth?

16 How on Earth?

Chris Forrester

A few hundred metres from Japan's public broadcaster NHK's giant Research & Development facility is Ultraman park, and an impressive statue to Ultraman, a 1960s TV futuristic superhero series that's remembered fondly by Japanese adults of a certain age. Indeed, the show was produced in a local studio and the district became known as UltraTown by residents, such was the fame of the show. NHK's nearby laboratory is nothing if not futuristic, with its 240 highly qualified researchers, of which 78 hold doctorates, and all focussed on tomorrow's TV.

Japanese public broadcaster NHK says it will start experimental test transmissions of its spectacular Ultra-HD system (7680 x 4320 pixels) in about 2012. Japan switches off its analogue terrestrial transmissions on July 24, 2011, and NHK will hold onto some of the frequencies permitting expansion into next-generation HDTV.

Kenkichi Tanioka, then Director General of NHK's science and technical research laboratories, was outlining NHK's timetable at a major presentation in Tokyo in 2008, saying that he anticipated further testing and development taking until 2016, and implementation then quickly following. "The biggest problem is bandwidth," he said, but praised the co-operation agreement now in place between the BBC, NAB, RAI and the EBU, which he described as a mutual approach to the evolving technology. He admitted that precisely predicting the future for Ultra-HD was especially difficult, "but we already make possible the impossible," he stated.

07/10/25
First Airbus A380 passenger flight, between Singapore and Sydney

07/11/05
Writers Guild of America strike starts, end February 12, 2008

C. Forrester (ed.), *High Above: The Untold story of Astra, Europe's leading satellite company*,
DOI 10.1007/978-3-642-12009-1_16, © Springer-Verlag Berlin Heidelberg 2011

Ultraman Jack poses with young fans

07/11/18
Sirius-4 launched from Baikonur

07/11/30
Evel Knievel, motorcycle stuntman, dies

07/12/27
Pakistani Prime Minister Benazir Bhutto assassinated

Unlike in the past when broadcasters from the USA, Europe and Japan fought each other over rival TV transmission systems (the USA's 'Grand Alliance', EBU's DVB and Japan's analogue Muse and then Hi-Vision digital system are all different), there seems to be a wish from all parties to avoid costly and overlapping effort in the search for next-generation HDTV.

NHK subsequently followed up these statements with spectacular live displays at IBC 2008, using London's South Bank near City Hall and Tower Bridge, with a backdrop of the City of London financial district.

The BBC supplied its Dirac advanced compression system into the transmissions. John Zubrzycki, the BBC's Principal Research Engineer at Kingswood Warren, says Dirac is achieving close to 1/200th compression, and a vital link in getting the massively fat 24 Gb/sec signals down to a more manageable 120Mb/sec.

Peter Wilson, Technical Advisor to the BBC's Research & Innovation division, and also present in Tokyo, said the BBC Dirac team had several directions to follow. "We are already standardising the technologies within SMPTE, and our original intent was to focus on professional applications, that is for extremely high-fidelity images within the broadcasting environment, for post-production and for transmission links and possibly storage. We also have a hardware partner called New Media Technologies with products in the market. But the other side of the coin was our wish to see Dirac used for transmission, including streaming. Here the emphasis is on bandwidth saving for transmission, but if we can achieve both, that

NHK expects projection to be one of the end uses for Ultra-HD throwing images onto a full wall of the living room or den, with a kids' corner sitting alongside the main screen area.

A Berlin experimental screening in 2010 of an impressive 180-degree, high-resolution display featuring 7 DLP projectors that deliver seamless images.

08/01/01
Cyprus and Malta adopt the euro

08/01/04
Tribal violence in Kenya, hundreds die

is bandwidth saving but also high-quality end results then we have a very appealing technology."

The London signals were piped to Amsterdam by Cable & Wireless. Kohji Mitani, a senior research engineer working on Ultra-HDTV, said the next stage is to start adding content for the broadcaster's own archive. "Co-operation is essential, not just between us engineers, but between engineering and production. Engineers do not usually make good producers and cameramen, nor cameramen good development engineers. We need creative co-operation."

His colleague Yudi Nojiri confirmed that huge progress was made in 2008 and 2009 with its 33 million-pixel camera. "We can now manage 10 times the sensitivity," said Nojiri. "Last year capturing twilight was hard, now it is possible." There are still obstacles to be overcome, not least in terms of studio use and a drive towards a single chip CCU (the camera currently uses 3 CCUs).

Then there's the display end of the transmission chain. Visitors to NHK can see showcased an extremely high-end 8K high-contrast projector passing light over a 33 million-pixel panel that the engineers claim delivers the whitest 'white' yet available from a projector, and the blackest black, and shades in between in a very wide dynamic range. While NHK expects projection to be one of the end uses for Ultra-HD throwing images onto a full wall of the living room or den, with a kids' corner sitting alongside the main screen area while at the same time perhaps showing data in an opposite corner. They also recognise that display panels will also be needed.

There are already 2K display models creeping onto the market, but there's some way to go to achieve 4K (7680 x 4320 pixels) super-fine resolution in terms of Plasma or LCD display, and the Holy Grail is to combine 0.3 mm-pixel pitch while at the same time reducing power demands to below that of a small power station! NHK demonstrated such a unit, with a claimed power savings of 30% over similar sized units – and is working on a 100-inch model yet with all the brilliance and luminosity needed for daylight viewing.

The prospects of Europe seeing full-time transmissions of Ultra-HD in the next 15 to 20 years must still be treated with caution. The question most often asked is "Do viewers need greater definition?", and it is a fair question to ask, especially if one accepts the fact that millions of homes have HD Ready displays but are – at best – watching just one or two HD channels and bypassing completely the pay TV, high-definition offerings. But it is also a fact that huge progress has been made in 2009 and 2010, and most experts recognise that the Consumer Electronics industry will be looking for the next 'big thing' probably less than ten years from now. In other words, never say "never" as far as improved resolution is concerned.

Besides, there are other developments that will affect us at home, and might well lead to additional demands on satellite bandwidth. The first refers to a natural expansion in the number of channels on offer, as well as a shift to HDTV. Paris-based EuroConsult's annual studies of the potential HD market just seem to grow their expected numbers every year.

In early 2009 it was estimated that 59 million European households were equipped with HD-enabled TV sets. That figure is set to grow to 116 million in 2010 – a 51% penetration rate and explode to 220 million in 2018, according to 'HDTV in Europe, Key Economics & Prospects to 2018', a recent report co-published by EuroConsult.
Driving this adoption is the steady decline in prices for flat screens since 2004, and the growing inclusion of HDTV reception as a standard feature in flat-panel screens sold throughout Europe.

08/02/08
Spain's first HDTV channel launched

08/02/27
SES orders NSS-14 from Space Systems/Loral

08/03/15
AMC-14 launched from Baikonur, but fails to reach orbit

08/02/18
Northern Rock bank nationalised in UK

08/02/19
President Fidel Castro of Cuba retires

08/03/14
Film director Anthony Minghella dies

California, perhaps the natural home of large-screen television sets, wants to ban big screens in order to save energy

A growing number of European households can receive high-definition content, another critical factor for market growth. The number of pay-TV networks in Europe offering HD almost tripled in the last two years. While satellite HD services still dominate HD subscriptions, cable and IPTV services are accelerating their introductions of HD offerings. And while satellite free-to-air may remain the only option in several countries in the short term, a major European 'first' occurred with the 2008 launch of a multichannel HDTV offer over digital terrestrial television (DTT) in France. Meanwhile, in 2008, the number of HD channels distributed in Europe more than doubled to 130 channels, and increased again in 2009. Over 600 HD channels should be distributed in Europe by 2013.

"Pay-TV reception will drive growth of HDTV adoption in Europe in the short term, despite the current economic downturn. In 2013, over 38 million households should receive HD pay-TV services, twice the number of households watching HD free-to-air only. From 56 million in 2013, the number of households receiving HD will boom to over 175 million households in 2018, making from HD the standard TV experience in most Western and Central European markets," indicated Pacôme Révillon, Managing Director at Euroconsult.

Incidentally, NHK is working hard with other local research & development projects. For example, Mitsubishi in 2009 released a giant – but prototype – 155-inch Organic Light Emitting Diode display, made up of 720 small 10-cm OLED blocks. This bright and low-power unit could revolutionise the way we view TV. OLED technology is based on organic materials which emit light naturally after an electrical charge is passed through them. OLED pixels generate their own light, which brings a whole host of technological advantages. Every OLED prototype we have seen produces brighter, sharper images while using less power than any plasma or LCD TV. Mitsubishi claims that its 155-inch OLED is around three times brighter than conventional LCD TVs. Sony, Samsung and LG are also working on similar technology.

NHK is also active in projector technology, seeking to bring the sort of spectacular results now available on high-end cinema displays, down to domestically affordable prices. The presence of small homes in Japan means that they need to solve the

08/03/18
83% of European TV homes are digital. Astra has 117 million homes

08/03/31
SES increases ownership of SES Sirius to 90%

08/04/08
SES Americom orders satellite from Orbital Sciences

08/03/19
Arthur C. Clarke, polymath, dies

08/03/24
Richard Widmark, actor dies

These tiny devices, about the size and shape of a 10-mm sugar cube, are designed to be integrated into cellular phones and beam images.

'ever larger' TV set problem, and see projector technology as providing the answer.

Related, and beginning to appear in the stores are PICO-cell projectors. These tiny devices, about the size and shape of a 10-mm sugar cube, are designed to be integrated into cellular phones and beam images, which might be a PowerPoint presentation or a TV signal, or cached video, onto a near-by wall. Battery life is a problem just at the moment, but they're working on this!

NHK's engineers are also looking at what can only be described as 'pen-TV', where a device that's shaped like a pen has a built in display screen that unfurls itself to a highly watchable (and in colour) flexible screen measuring about 10cm x 4cm – much larger than anything a cell phone can offer. The

Japanese electronics giant Mitsubishi Electric displays a 155-inch sized organic light emitting diode (OLED) display

08/04/11
AMC-14 declared a total loss

08/04/29
Astra opens new position at 31.5 degrees East

08/04/05
Charlton Heston, actor, dies

08/04/28
India sets world record, ten satellites on a single launch

battery and the receiving equipment are housed in the body of the 'pen' which also acts as an aerial.

But as we have often had to say during these pages, there has to be a 'health warning' on any of this Crystal Ball gazing. Our national press, and even specialised media press, are filled with spectacular forecasts of how the future might turn out. And remember the most famous family of the future, The Jetsons, was produced by Hanna-Barbera's animation studios back in 1962, and they got things badly wrong!

At an Astra conference in 2006, delegates heard from Professor Peter Cochrane OBE, head of ConceptLabs, but formerly Chief Technologist at British Telecom's Research & Development division. He warned about looking too far into the future – and attempting to predict the future. He had another snappy word for such predictions: he called it "guessing". He's right. He explains that any development likely to be introduced in the next five years is already in a laboratory test-bench now. If it isn't on that test-bench then it is likely to be ten years away, if ever!

Cochrane in June 2009 pointed to there being over four billion mobile phones worldwide, with over 50% of those being Internet-ready; and mentioned his Linked-In account, which shows that there are already 7,982,000 people just three degrees away from him. He said that in 2006, the Internet had reached the same number of nodes as a human brain; by 2012, it would have a thousand times the number of nodes; and by 2018, a million times the number of nodes. He says that laptops aren't smart: they have no sensors and no concept of adaptability; but that as we build more sensors into things, our machines will be capable of being much smarter. He points to the clever iPhone, with its camera, its compass, its accelerometer and a zillion 'apps'. He says that humans can't get more intelligent, while computers can.

They can access more data, and extract knowledge more easily. He points to a future which is more run by computers than ever before.

This is perhaps key. We human beings can only change a little. A new device might be fascinating, and best learned by a 12-year-old. Those of us who are a little older sometimes find it tough to take on new technology skills. While entertainment is mostly a passive exercise and should be created by humans, the best overall advice for humanity is to 'stay flexible' (and healthy).

This is the accepted mantra of the satellite industry. Satellites are designed to be flexible (and healthy), thus being ready at the press of a button to take on a new, or developed role. Astra 1A is long gone into the great satellite graveyard in the sky, but if it were still around the fact is that it would be able to handle HDTV signals (or 3D) just as well as the latest Astra satellite. That's flexibility, and the wonderful power of satellite.

Any development likely to be introduced in the next five years is already in a laboratory test-bench now. If it isn't on that test-bench then it is likely to be ten years away, if ever.

An Astra satellite in final testing

08/07/14
SES orders Astra 1N, from Astrium, for delivery in 2011

08/08/15
Americom's AMC-21 launched from Kourou

08/07/02
Íngrid Betancourt and 14 other hostages rescued

08/08/08
Summer Olympics start in Beijing

08/09/15
Lehman Brothers files for bankruptcy protection

17

Hard Science or Science Fiction?

17
Hard Science or Science Fiction?

Chris Forrester

A new decade, and a special period for Astra as it embarks on its 26th year, gives every press and broadcasting pundit the opportunity to sound off on his or her pet futurology theories. There are, however, some voices which should be listened to. The first set comes from the cutting edge of the business marketplace – a global business that is well known for its sharp appreciation of customer trends and developments.

Pricewaterhouse Coopers, in a report some ten years ago that they themselves called 'A Gaze into the Crystal Ball' spoke optimistically about our digital future. Report author and digital pundit Saul Berman says the Digital Age of television is upon us. "3D, 360-degree audiovisual experiences are all the rage at theme parks...and the Internet has spawned a vital new home information, education and entertainment paradigm."

While no one can know for sure what the frenetically evolving consumer entertainment landscape will look like as we get deeper into the 21st century, Berman's views of how his crystal ball is appearing makes interesting reading in the context of this book. However, it is also an object lesson in that we should not expect too much too soon.

It is now clear that what Berman called 3D television is very much a reality. BSkyB will launch a dedicated 3D channel in 2010. Major broadcasters such as ESPN and Discovery Networks are also very active, as is DirecTV in the USA. Others will surely follow, although it is going to be an expensive hobby. David Hill, President of Fox Sports in the USA, and an old BSkyB hand, warned IBC delegates at the 2009 event not to expect too much from network TV as regards 3D.

He expressed himself as a fan of 3D-TV technology, but as a broadcaster is not happy with the economics. "Let me tell you, the

08/10/08
SES orders new Sirius 5 craft from Space Systems/Loral

08/10/08
Astra HD channels grow by 30% in three months

08/09/26
Paul Newman, actor, dies

08/10/21
CERN's Large Hadron Collider (LHC) inaugurated

C. Forrester (ed.), *High Above: The Untold story of Astra, Europe's leading satellite company*, DOI 10.1007/978-3-642-12009-1_17, © Springer-Verlag Berlin Heidelberg 2011

broadcasting industry will not be investing in 3D technology until we know if we are going to be reimbursed one way or the other."

Considered the doyen of sports broadcasting, Hill is a voice that the industry listens to. At the moment he is not happy. "As a viewer I think 3D is fabulous, but as an executive I hate it because we still haven't finished paying for HDTV," says Hill. He says the mandated move to HD in the USA was "one of the greatest tragedies of our times" because the economics of HD for broadcasters don't add up. "The set manufacturers, the equipment manufacturers, [they] all made money off of HD and I am still having to put out millions and millions of dollars, which means I can't do programme innovation," Hill claims.

"I can't give people raises, I can't employ any more people. I can't do the stuff on the Internet that I want," complains Hill. "Why? Because there is this big truckload of money departing to the Sonys and the Panavisions and the people who make the equipment and the wires for HD."

Hill showed a stunning 3D tape of sports at his keynote at IBC 2009 that included boxing and football. The problem is that he can't ask the advertisers to pay extra for these improved TV pictures. "What are advertisers going to do if I ask them to pay double for 3D pictures? They are going to say this," he said, making what can best be described as a universal gesture of disapproval. Underlining his point, he added: "We are not a public service broadcaster. We are in this for the dough," Hill said.

"3D is going to be fantastic, but only if the sums can be done."

"3D is a case of fool me once, shame on you, fool me twice, shame on me," said Hill. "Let me tell you, the broadcasting industry will not be investing in 3D technology until we

Fine arts: BSkyB is capturing more than just soccer 3D content

08/11/05
Astra 1M launched from Baikonur

08/11/25
SES and EchoStar to expand DTH over Mexico/USA

08/12/08
Ciel II launched from Baikonur

08/11/04
Barack Obama elected 44th President of the USA

08/11/26
Mumbai is hit by ten coordinated terror attacks. 173 die.

08/12/12
Switzerland joins Schengen Agreement

know if we are going to be reimbursed one way or the other. I am only talking about the part of the broadcasting industry that I control but if the others have any sense they will say exactly the same thing."

"We are still getting over the cost of HD. It cost us a fortune and not just in our trucks but in our internal distribution and infrastructure," says Hill. "A bunch of people made money on HD, but not us." He says that he is not planning on further upgrading his studio to HD because it will cost 18 million dollars. "I'm not going to do it."

Hill gave three criteria that are needed for 3D to work for broadcasters: a common standard, a way to recompense broadcasters for the new financial outlay and making the 3D broadcast system technically 'watertight'. Hill admits he doesn't know what the revenue model is for 3D: "I want someone to come into my office and write a bloody big cheque," he said. But he suspected one revenue stream is going to be around the 3D glasses themselves. "Tommy Hilfiger is going to make some cool 3D glasses so that when a guy asks a girl out on a date he's not wearing these iron-welding-looking glasses. That's at least one revenue model."

2009

09/01/20
Astra 1M operational

09/01/15
US Airways plane lands in Hudson River

09/03/16
Astra supplies payload for Egnos navigation service

09/02/07
Deadly Australian bushfires

David against Goliath, rising start-ups and struggling stars: The big question that remains is who will take us to this broadband promised land, providing us with the fat communications pipe to make our 21st Century dreams a reality.

The wood is cluttered with so many trees...

The rather sudden emergence of 3D might have come as a surprise to some even hard-boiled industry experts. Besides all the usual predictions about television becoming more interactive, and suggesting that everyone would become his own programming director, as well as continuing diversification, programming proliferation and the possible commoditisation of the distribution capacities, the rise of HD and 3D rather proves that while the consumer remains all-powerful the linear TV industry is not at risk. We have a fundamental need for mass entertainment and despite multiple choices this basic need has not changed that much over the past ten years. Indeed, in most markets TV viewing is up, not down. TV viewing is an incredible 219 minutes average viewing time per day in Europe, 55 days per year, and television continues to maintain its social function and strengths. Community websites show that sharing of content remains a major need and expectation from media. It might be easily argued that the complete individualisation of TV is a myth, at least for the foreseeable future.

Of course, there will be complementary entertainment, there will be substitutions. Interactivity is playing its roles, and TV is definitely becoming more mobile. The likes of CNN, the BBC, Al Jazeera or Japan's NHK, are all now streaming to cell phones as well as the Web. Consumers seem ever ready to consume more hard news, whether stocks and shares, sports or just a missed recipe from yesterday's cookery show! Chef Jamie Oliver's cellphone recipes, a direct spin-off from his hugely popular TV show, were one of the iPhone's big download 'hits' of 2009, and at 6 euros to view.

The crisis in the world's advertising markets forced television companies to look for alternative revenue sources, the pressure on cost and margins undoubtedly also has a (hopefully only interim) negative impact on programming quality. Alternative business models are urgently needed, and broadcasters are looking for more time for commercial messages, on-screen product placement, and spin-off speciality channels that can earn extra cash. Hollywood, and sports producers, will not be slow to capitalise on fresh premium content, whether movies or sports games in 3D, as well as a myriad new ways to slice up library content.

New players are entering the field, creating exciting new alliances, spanning from major players in the consumers electronics and gaming businesses to other internet, high-tech and telecommunications companies. Broadcasters are afraid of the threats from those with deep pockets, such as the telcos who are entering internet TV – and at the same time reassure themselves that these attacks are only a defence for their ailing core fixed line business and that they would come under massive pressure to finance their further broadband extension and their costly content adventures.

David against Goliath, rising start-ups and struggling stars: The big question that remains is who will take us to this broadband promised land, providing us with the fat communications pipe to make our 21st Century dreams a reality?

The answer is certainly that satellite will keep its major role in this. The technological and therefore economic advantages of satellite for broadcasting – and which has allowed its successful development over the last 25 years – will remain unchanged, strong and extremely influential. The broadest variety of programming in the highest possible quality and the largest geographical reach: no other infrastructure will – without massive investment and a lot of time – be able to replace that. DTH therefore is an extremely stable business model. TV creates mass audiences. Satellite is key for its distribution and will continue to prove to be flexible and innovative enough to reply to these substitutive threats.

09/03/24
Astra's European reach now 122 million homes

09/04/06
Solaris Mobile launched

09/04/20
Astra and YahSat form Middle East partnership

09/03/03
Sri Lankan cricket team attacked at Lahore, Pakistan

09/04/06
6.3 earthquake hits L'Aquila, Italy. Almost 300 die

09/06/25
Farrah Fawcett, Charley's Angel, dies

Our 'Top Five' predictions for digital television and its next stages

BIG CHANGE 01
3D and even Higher Definition

We have covered the growing likelihood that 3D television will have on our viewing experiences. Funding 3D is relatively straightforward for pay-TV operators. They present a number of channels to subscribers and see how well they do in the marketplace! But public or commercial networks must somehow square a very difficult circle: Advertisers seem reluctant to pay for extra HDTV quality, let alone 3D transmissions, and finding the extra production costs for 3D might be difficult. However, there's an immense supply-line 'push' for 3D from the manufacturing side of the business. It will be interesting to see how this translates into consumer 'pull'. There's also the vexed question of transmission standards to be resolved.

However, there's one element of higher definition that seems inevitable, and that's the drive toward ever higher transmission image quality. The past few years have seen the production community and broadcasters embrace HDTV. Suitable set ownership is widespread but there's still an element of education needed if all the 'HD Ready' displays are really to start watching true high definition.

The past few years have also seen broadcasters move from 720i to 1080i to 1080p/25Hz and in some markets transmitting in 1080p at 50/60Hz and thus a truly impressive HD image. But where does the industry go from here?

In fact, this element alone could occupy SES Astra for the next 25 years! The Japanese are hugely enthusiastic about 4,000 line (and even 8,000 line) image capture and transmission. Given that the TV display industry is dominated by Japanese and South Korean manufacturers, it does seem inevitable that the marriage between Japan's broadcasting sector and the display industry will continue to push image boundaries to ever higher levels of quality – and there's one inevitable fact of life in all this Crystal Ball gazing, and that image quality soaks up bandwidth, and satellite is best able to satisfy this demand.

BIG CHANGE 02
Personal Video Recorders

Put any two so-called experts together and you will have two different views of where the next television revolution will come. But few doubt that one key development is the new generation of set-top boxes now quickly emerging. In essence, they allow viewers to throw away the old video recorder. The new PVRs (Personal Video Recorders) have computer-type hard drives that are growing in size, and Terabyte-size storage is increasingly common. Storage will get larger, and the receivers themselves more sophisticated. This has the end result of placing a high-capacity file server in every home.

These PVR set-top boxes, as well as picking up satellite/cable signals, also allow viewers to record programmes, just like the VCR. But the key difference is the 'intelligence' built into the PVRs. This 'fuzzy logic' intelligence monitors a user's viewing habits, and can record a series, or multiple series automatically at the push of a button or even with remote commands. Units with five or more tuners are increasingly commonplace, and as if these were not sufficient, there's a trend towards more! The logic behind this move is the 'home network' role, where the single unit might have to 'receive' and 'send' programming to slave units elsewhere in a home, perhaps to bedrooms, a family den or kitchen, as well as the 'main' set.

BIG CHANGE 03 ADSL

Another favoured option is ADSL (Asymmetrical Digital Subscriber Line) and/or IP-based television technology designed to be delivered by telephone or cable companies. Utilising digital technology, telephone companies can supply almost any number of channels down an 'ordinary' telephone line. In fact, the line becomes two-way, transforming a phone line into a reasonably fast, two-way pipe capable of running video, interactive games and TV on demand. The line will have been significantly upgraded, vital if bothersome (and technically disastrous) pops, clicks and static stay on the system. The line is also enhanced at the telephone exchange, with the telephone company supplying a choice of channels just as if it were a satellite or cable company. These ADSL/IPTV technologies are now being widely implemented.

BIG CHANGE 04 Webcasting

'Television' is already widely available on the Internet, and viewers are increasingly seeing the power of using Web-based technology to marry Internet-delivered TV signals onto their computers. While it is generally accepted that the Personal Computer is being used for 'lean forward' activity, while the TV set is 'lean back' or relaxation mode, it is also a fact that the lines are being blurred, with TV increasingly offering niche 'teleshopping' options, and computers happily streaming video signals.

But there's another development that places even greater interactivity into the TV set. Leading TV manufacturers are adding Ethernet and WiFi-type functionality directly into their TV sets. They are already supplying set-based 'widgets' to enable web-surfing, and while the primary objective is simply to gain access to services like YouTube, there is no reason why extra choices TV should not be available. Some pundits have even suggested that giant media players like Sony, already making movies, music, Blu-ray units and TV sets, might even enter the pay-TV field via these web-based hook-ups.

BIG CHANGE 05 Compression

Some say that more channels means dramatically less expensive transmission charges, and as a finance director of a broadcasting station you would like to believe in this. However, the proliferation of digital offers and the emergence of HD and 3D mean, at the same time, higher and higher bandwidth demands from traditional broadcasters as well as emerging telcos and therefore lead satellite companies to optimistic demand forecasts for the near and mid-term future. This will be partially offset, of course, by the move from MPEG-2 to MPEG-4 (similar to the one from analogue to digital transmission), as a higher compression standard, allowing to make a more efficient use of the rented capacity. But the future of satellite transmission and transponders remains bright. The rising interest of not-broadcasting costumers like governments will further contribute to the outlook. Satellite seems to be here to stay.

09/10/29
NSS-12 launched from Kourou. Sky Deutschland adds HDTV channels

09/11/01
Launch of HD+ platform in Germany

09/11/30
SES orders four satellites from Astrium

09/10/02
IOC awards 2016 Summer Olympics to Rio de Janeiro

09/12/10
James Cameron's 'Avatar' premieres in London

09/12/18
Copenhagen climate summit

Our Final 'Health Warning'

In talking of these 'big changes' in digital television, I am reminded of a conversation I had with a senior executive of a UK television rental company (for non-UK readers, the UK once had a tradition of people renting, not buying, their television sets) who reminded me that they still had more than 20,000 people paying for 'black and white' television reception, this a few generations after the introduction of colour TV. Indeed, these customers had to be issued with colour sets that had been deliberately degraded to keep them within the government regulations on colour TV reception (the UK government allows a licence fee discount for monochrome reception).

I make the point if only to remind ourselves that for every dynamic early adopter of these technologies there will often be a matching laggard, and as Europe moves towards totally embracing digital broadcasting, not all viewers will be enthusiastic about a multi-channel future promising dozens or hundreds, let alone thousands of potential channels. Politicians everywhere will have to face the electoral consequences of balancing their enthusiasm for these new technologies while at the same time paying due regard to those members of the population who are less than enthusiastic for such social changes.

2010

10/01/12
Astra awarded second Egnos contract

10/01/04
In Dubai, the Burj Khalifa opens, the world's tallest structure

10/01/01
ProSiebenSat.1 joins HD+ platform in Germany

10/01/12
Catastrophic earthquake hits Haiti

10/03/07
The 'Hurt Locker' wins 6 Oscars

18

The Astra Chronicles

18 The Astra Chronicles

Some key historical dates in broadcasting and satellite

Some of our very early dates in television have to be treated cautiously. Patent dates have been given, or described wherever possible, although exhibition dates for a device might be a year or more after the patent filing.

C. Forrester (ed.), *High Above: The Untold story of Astra, Europe's leading satellite company*,
DOI 10.1007/978-3-642-12009-1_18,

1878
In a May 1878 edition of Scientific American, the publication referred to the work of George R. Carey, a professional surveyor working for the city of Boston, in using selenium photocells (and wires) to transmit a signal. The end result was a series of individual wired 'pixels' that could be viewed. The following year, the Scientific American reported in detail on the invention, describing it as "seeing by electricity" and talking about Carey's "selenium camera".

1884
Paul Nipkow unveils a scanning disk television system (his "electric telescope"). He patented the concept in 1878.

1887
In 1887 Hertz, a German physicist demonstrated how an electromagnetic wave (itself discovered by Scotsman James Clerk Maxwell) could be made to jump in the form of a spark across an air gap.

1888
The German physicist Heinrich Hertz generates electromagnetic waves and thus creates the prerequisite for the later broadcasting technology.

1897
German physicist Karl Ferdinand Braun invents the Cathode Ray tube. He received the Nobel Prize (for physics) in 1909.

1900
At the Paris World's Fair, the first International Congress of Electricity is held. That is where Russian Constantin Perskyi made the first known use of the word 'television'.

1901
Guglielmo Marconi sends a transoceanic radio signal from England to Signal Hill, Newfoundland. In the same year, Marconi patents a selective tuning device for receivers. In 1897 he had received his first British patent for wireless telegraphy.

1904
J. Ambrose Fleming patents the Fleming vacuum tube 'valve'; Lee deForest builds on the concept and introduces the Audion amplifier tube. DeForest's patent is granted on January 15, 1907.

1906
Canadian inventor Reginald Fessenden demonstrates wireless telephony, a means for radio waves to carry signals a significant distance. He was working at the U.S. Weather Bureau at the time.

1907
Russian scientist Boris Rosing combines Nipkow's disk and a cathode ray tube and builds the first working mechanical TV system. He files a patent on November 26, 1907. Vladimir Zworykin (see 1923) was a pupil of Rosing.

1920
KDKA, a Pittsburgh Westinghouse station, transmits the first commercial radio broadcast.

1922
British Broadcasting Company is formed. First broadcasts from station '2LO' on November 14.

-

Founding of the first German broadcasting company in Berlin.

1923
Russian immigrant Vladimir Zworykin patents the iconoscope, the first television transmission tube. He patents the first colour tube in 1925.

-

A.C. Nielsen Company is founded and provides measurements of radio audiences for advertisers in the USA.

-

First public radio transmission from VOX in Berlin.

1925
John Logie Baird demonstrates his motion system at the Selfridges department store in London, on March 25.

-

In May, AT&T's Bell Telephone Laboratories transmits half-tone (but still) images.

-

On June 13 Charles Francis Jenkins transmits a silhouette moving image of a toy windmill over a distance of five miles. It has a 48-line resolution.

-

On October 25 Baird manages to transmit moving images, but at just five frames per second.

1926
John Logie Baird launches his 'televisor' onto the UK. By January 1926, Baird manages to transmit images at 12.5 frames per second using 30 lines.

-

RCA, General Electric and Westinghouse establish NBC, which operates two national radio networks (the RED network, and the BLUE network) in the USA. NBC Blue is eventually turned into ABC in 1945.

-

On December 25 Kenjiro Takayanagi demonstrates a 40-line TV system using a Nipkow disc scanner (see 1884) in Japan. By 1927 he manages 100 lines, a resolution not beaten until RCA in 1931.

1927
Philo T. Farnsworth transmits the first all-electronic television image. The "image" measures about the size of a postage stamp. He receives a patent in 1930. In 1928 Farnsworth demonstrates his electronic system to the press.

-

British Broadcasting Corporation is formed (see also 1922) and is granted a Royal Charter to operate.

1928

John Logie Baird beams a mechanical television image from England to the United States.

-

General Electric introduces a television set with a 3 inch × 4 inch screen.

-

The first television in the USA is sold – a Daven for 75 dollars.

-

In Pula, in what is now Croatia (but was then part of the Austria-Hungary empire), Herman Potočnik writes 'The Problem of Space Travel' in which he carefully outlines the concept of geostationary space stations and first calculates the geostationary orbit. He also outlines how they could be used for radio communications.

1929

CBS in the USA is founded by William S. Paley.

-

John Logie Baird opens the first TV studio, although the 30-line image was said to be 'poor'.

1930

The BBC in London begins test TV experimental transmissions.

-

German scientist Fritz Schroeter applies for a patent for interlaced scanning.

1931

John Logie Baird demonstrates his 'Zone Television' in which people can be seen (as distinct to heads/shoulders). A lesson in how to play cricket is shown.

1932

John Logie Baird supervises the showing of a programme from Broadcasting House, London, to the Arena Theatre in Copenhagen, Denmark, some 600 miles away.

1933

Edwin Armstrong introduces Frequency Modulation (FM), a static-free method of transmission.

1934

Philo T. Farnsworth demonstrates an all-electronic complete TV system at the Franklin Institute in Philadelphia.

1935

AEG/Telefunken exhibits the first magnetic tape recorder in Germany.

-

On February 13, 1935, France inaugurates a 60-line system that broadcasts for 15 minutes each evening. Broadcasts ramped up from Jan 4, 1936 when three separate system experiments also took place.

1936

The BBC begins 'high-definition' broadcasts from London – three hours a day initially using Baird's 240-line system although the image was captured on 35mm film (described as the intermediate film system) and then transmitted. This system was dropped after three months in favour of EMI-Marconi's 405-line, all-electronic 'Emitron' camera system (and thus 'high-definition' at the time). The BBC signals were the world's first regular high-definition television service. The Emitron camera was designed by Isaac Shoenberg.

-

November 10, 1936 saw a 180-line experimental TV service being beamed from the Eiffel Tower to the citizens of the Paris region.

1937
Edgar Bergen and his puppet Charlie McCarthy, make their TV debut on NBC.

1938
First UK public demonstration of large-screen colour television at London's Dominion Theatre by John Logie Baird and transmitted from the Baird studio at Crystal Palace in South London.

-

In France in July 1938, a decree defines for three years a standard of 455 lines VHF (whereas three standards were used for the earlier experiments: 441 lines for Gramont, 450 lines for the Compagnie des Compteurs and 455 for Thomson). In 1939, there were about only 200 to 300 individual television sets in use, most in retail stores and other public locations.

1939
Electronic television showcased at the New York World's Fair. RCA/NBC uses 500 horizontal lines. DuMont, a rival TV broadcaster, starts transmissions using 750 lines, and sells its own TV sets.

-

On May 13, 1939 the first public demonstration of TV in Japan takes place. A signal was transmitted 13km to Tokyo's Broadcasting Hall.

-

First Italian test TV transmissions.

1940
CBS demonstrates colour television in New York.

1941
The USA's National Television Systems Committee adopts its NTSC Standard of 525 interlaced lines.

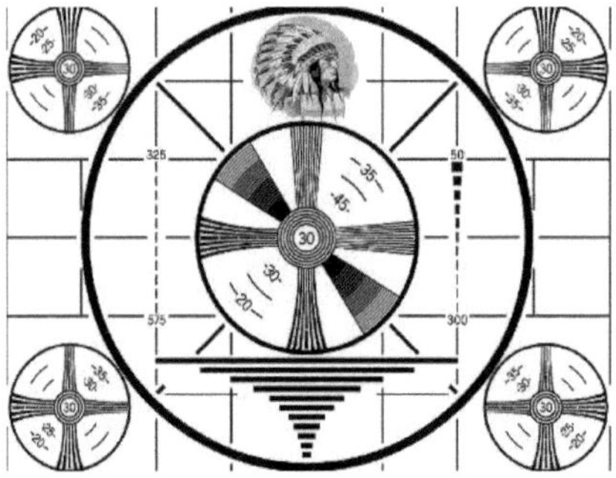

1944
TV broadcasting resumes in France following the end of WWII.

1945
The British mathematician and physicist Arthur C. Clarke, in a technical journal, says that a satellite positioned at an altitude of 35,803 km above the equator needs exactly 24 hours for one revolution.

1946
Peter Goldmark, working for CBS, demonstrates a colour TV system but using a spinning red-blue-green wheel.

-

The world's first 'television sports extravaganza' is shown in June, when a boxing match between Joe Louis and Billy Conn for the heavyweight championship takes place. There is a claimed TV audience of 150,000, watching on an estimated 5,000 TV sets (30 people per set). But the event helps propel sales.

1947
The AT&T Bell Laboratories invents the transistor, leading to smaller, more efficient electrical circuits.

1948
Cable TV is introduced in Pennsylvania and Oregon (USA).

-

1 million TV sets in use in USA.

-

The US authorities licence 108 stations this year.

1950
First cross-channel TV transmissions take place between England and France.

-

German public broadcaster ARD is founded.

1951
Colour TV is introduced in the USA by CBS. Its transmission methodology is altered in 1953.

-

The first television test image is broadcast in Germany.

-

For the first time, a US nationwide programme airs. Edward R. Murrow, in the first broadcast of his 'See It Now' series, tells viewers, as he looks into the split-screen image of the Golden Gate and Brooklyn bridges, that they are able to see the Atlantic and Pacific oceans for the first time simultaneously.

1952

Official start of television in the Federal Republic of Germany.

-

Official start of Canadian broadcasting.

1953

FCC approves RCA's colour TV system, and rescinds a 1950 decision on the rival CBS colour system.

-

Queen Elizabeth II is crowned in Westminster Abbey, London.

-

Belgium starts TV transmissions.

1954

The collaborative programme of German television goes into service on November 1 (initially with six stations).

1955

The 100,000th television set is sold in Germany.

-

ITV, the UK's second channel, is launched.

-

Official start of television in the German Democratic Republic on December 21.

-

Luxembourg and Finland both start TV transmissions.

1956

Ampex (earlier founded by A. M Poliakoff, who added the EX "for excellence") demonstrates the first practical videotape recorder in Redwood City, CA. The early team included Ray Dolby, who later went on to found his famous 'noise reduction' company (Dolby Labs).

-

Start of commercial (non-public) television in Germany.

-

The Zenith Space Commander, the world's first wireless TV remote control, is introduced (invented by Austrian Robert Adler).

-

The first Eurovision Song Contest is transmitted.

-

Australia, Romania, Algeria, Austria, Iraq, South Korea, Spain, Sweden, Uruguay, Yugoslavia and East Germany all start regular TV transmissions. The catalyst for many of these new services was that year's Olympic Games, from Melbourne, Australia.

1957

Sputnik 1, first man-made satellite of the USSR, orbits the Earth for 96 minutes.

1958

Explorer 1, the first American satellite, arrives in space.

-

Founding of the space agency NASA.

-

The USA has 525 cable TV systems running, serving 450,000 homes. CBS runs an advertisement saying that "free TV cannot survive alongside pay TV".

NASA

1959

India starts TV broadcasting.

1960

'Tiros 1', the first NASA weather satellite, is launched. It broadcasts television pictures of a tropical hurricane in the South Pacific before the storm's existence is known.

-

Echo 1, the first passive news satellite, launches. It is a spherical balloon with a diameter of 40 meters and an aluminised shell, which reflects radio and television signals to Earth.

-

Norway starts TV transmissions.

-

Egypt and Syria start broadcasting.

1961

The Soviet cosmonaut Yuri Gagarin orbits the Earth as the first person in outer space.

1962
Telstar 1, the first commercial television satellite, is launched. The first direct television broadcast between the USA, Europe and Japan takes place and reaches 200 million viewers.
-
The first live programme is transmitted via the Eurovision Network.

1963
ZDF starts as the station of the German federal states.

1964
In Germany, the 10,000,000th TV set is plugged in.
-
Eleven countries sign the charter of the International Telecommunications Satellite Organization (Intelsat).
-
'Peyton Place' premieres on ABC in the USA and is the first prime time soap opera.
-
BBC2, the BBC's second channel, launches.

1965
The first commercial news satellite launches – Early Bird (Intelsat-1). It weighs 409 kilograms and has a capacity of 240 telephone channels or 1 television channel.
-
Ghana, Paraguay, Senegal and South Vietnam start TV broadcasting.

1966
Greece, Tunisia, Iceland, Israel and Cambodia all start TV.

1967
The Intelsat-2 Satellite makes multi-point communication connections between Earth receiving stations within its footprint possible for the first time.

1968
NASA is able to make the first manned lunar orbit with the space shuttle Apollo 8. This major event is broadcast live on television.
-
Television programmes and telephone conversations or data transfer can be transferred at the same time via Intelsat-3.
-
Jordan starts broadcasting, as do Turkey, Equatorial Guinea and Libya.

1969
The American astronaut Neil Armstrong is the first person to set foot on the moon at 3.56 CET. An estimated 600 million viewers tune in to see the transmission.
-
Intelsat-4 launched. It has 4,000 telephone/data lines and two television channels with 12 transponders.
-
Public Service Broadcasting starts in the USA. In November, Sesame Street is first aired on PBS.
-
The United Arab Emirates starts TV transmissions.

1970
Coca-Cola's 'I'd like to teach the world to sing' commercial airs.
-
Qatar starts TV transmissions.

1973
Bahrain starts TV broadcasting.

1975
Home Box Office, a pay-TV channel, is launched in the USA.
-
RCA American Communications is founded as operator of the RCA Astro satellites. Satcom 1, its first satellite, is launched on Dec 12, 1975. This satellite helped pioneer the launch of channels like Ted Turner's Superstation, and CBN. The satellite had 24 transponders, and could thus carry 24 channels.

1976
South Africa starts TV broadcasting.

1977

The WARC (World Administrative Radio Conference), based in Geneva, decides on a worldwide broadcasting satellite plan. Every country can emit five TV-programmes (or several broadcast stations) directly from the satellite to the subscriber. The DBS-satellites (Direct Broadcast Satellites) will be positioned at an altitude of 36,000 km and with a separation distance of 6 degrees.

-

The first European meteorological satellite is called Meteosat 1.

-

Founding of the European Telecommunications Satellite Organisation Eutelsat.

-

'Roots' is broadcast in the USA, and draws an audience of 130 million viewers.

1978

Sky Television, a single satellite channel, is created.

-

Eutelsat sends its Orbital Test Satellite (OTS) into outer space.

-

Sony introduces its Betamax video cassette recorder.

-

Viacom launches Showtime, a pay-TV channel, in the USA.

1979

First flight of the Ariane rocket. The European Space Agency ESA starts its first unmanned launch vehicle from the European space port Kourou (French Guyana). Ariane is a three-stage carrier rocket, which has a transport capacity of two satellites with a total weight of 4 tonnes.

-

ESPN launches in the USA.

1980

Ted Turner launches CNN, the first all-news cable network.

-

Signing of the agreement between Germany and France on the development and testing of direct broadcast satellites.

-

'Who shot JR?' episode of Dallas airs.

-

MTV is launched in August.

1981

The French-German consortium Eurosatellit begins construction of the TV Sat 1 for the Federal Republic of Germany and TDF-1 for France.

-

NHK, Japan's public broadcaster, demonstrates its MUSE high-definition (analogue) TV system. It has 1,125 lines.

-

Prince Charles and Lady Diana marry.

1982

Telefunken presents the prototype of a receiver for Digital Satellite Radio (DSR).

-

Noise-reduction specialist Dolby introduces its Surround Sound system for TV.

-

The UK's Channel 4 launches.

-

Rupert Murdoch buys Sky Television (see 1978).

-

The German Federal Post decides to assemble a national telecommunication satellite system.

1983

A German industrial consortium receives the order to build a turnkey satellite system (DFS Kopernikus).

-

The first ECS-Satellite (European Communication Satellite) is launched into orbit.

-

Final episode of 'M*A*S*H' airs in the USA, and gains the largest audience in US television history, of 125 million homes tuning in.

1984

SAT.1, in which numerous newspaper and magazine publishing companies have a share, goes on air.

-

RTL plus (later simply RTL) starts its German-speaking TV programme.

-

USA approves TV stereo sound system.

1985

The Société Européenne des Satellites (SES) is founded in Luxembourg. Major European banks as well as the state Luxembourg are shareholders. SES is the first private operating company in Europe to be granted a concession to use satellites for an initial 22 years.

1986

The Space Shuttle 'Challenger' explodes shortly after launch.

-

An EC Directive from 1986 obliges existing satellites to transmit in the new standard D2-Mac.

-

UK allocates its broadcast satellite licences to British Satellite Broadcasting, BSB.

-

General Electric acquires RCA, and renames the business GE American Communications.

1987

Eutelsat's first satellite ECS 4 is successfully brought into orbit with Ariane 3 and put into operation.

-

The start of Germany's first direct broadcast satellite Sat1 is thwarted: One of the two solar collectors cannot be deployed, which prevents it being put into service.

-

More than 50% of US homes are wired to cable TV systems.

-

The broadcasting treaty, passed by the German federal states, offers unlimited reception of internal and external satellite programmes via parabolic antenna and cable connection.

1988

The first Astra satellite 1A is launched into outer space by the European carrier rocket Ariane and brought into orbit at position 19.2 degrees East. It enables the start of the private Astra system for the broadcasting of television and radio programmes.

-

In June, Rupert Murdoch announces the upcoming launch of Sky TV.

-

98% of US homes have at least one television set.

-

Ted Turner starts Turner Network Television (TNT) and buys MGM's film library.

1989

Astra 1A starts transmissions. News Corp's Sky Television takes a number of transponders. SAT.1, Pro 7, teleclub and RTL plus also book transponders.

-

The German communications satellite DFS Kopernikus is positioned at 23.5 degrees East.

-

Time Inc. and Warner Communications merge.

-

The television channel Eureka TV turns into Pro 7.

-

Fall of the Berlin Wall.

1990

The European Court of Human Rights in Strasbourg renders the so-called 'Autronic Ruling': The freedom of reception of broadcasting programmes from Article 10 of the European Human Rights Convention is subsequently protected. Everyone has freedom of expression. This also includes the reception and the communication of news or ideas without regard to national boundaries.

-

British Satellite Broadcasting (BSB) starts transmitting (April). By November its losses were so great that it agreed to 'merge' with Sky Television, thereby creating BSkyB.

-

'The Simpsons' debuts on Fox in the USA and becomes an instant hit.

-

'Seinfeld' debuts on NBC.

1991

Astra 1B is launched (on March 2), adding another 16 channels to its portfolio. A Lockheed-Martin craft, it is more powerful than Astra 1A.

-

Pay TV channel Premiere goes on the air in Germany.

-

The German Federal Ministry of Post and Telecommunication does away with the individual authorization of dish antennas for the reception of radio and television programmes, which was subject to fees.

The central facilities of GDR television are dissolved on December 31.

-

Persian Gulf War sees extensive coverage by CNN's signals which go around the world.

1992

There are an estimated 900 million television sets in use around the world; 201 million are in the United States.

-

Johnny Carson hosts 'The Tonight Show' for the last time. He had ruled late-night television for 20 years.

1993

Deutsches Sportfernsehen (formerly Tele 5), VOX, RTL 2 and VIVA go on air.

-

With the launch (May 12, 1993) of the third Astra satellite 1C, SES succeeds in making a world premiere: the co-location of 3 satellites (Astra 1A, Astra 1B and Astra 1C) on the same orbit (19.2 degrees East). The reception of 50 TV programmes is now possible.

-

ZDF decides to broadcast its main programme via Astra.

-

The final broadcast of NBC's 11-year hit show 'Cheers' airs. 93 million viewers tune in.

-

Time Warner launches its 'Full Service Network' in Orlando, Florida.

1994

Astra 1D is launched. Manufactured by Hughes (now Boeing Satellite), it is an HS601 tri-axis craft, and carries 24 transponders – 18 as primary and six as back-up.

-

Deutsche Telekom, the biggest competitor for the Astra satellite operators, becomes the largest shareholder in SES with 20%.

-

O.J. Simpson is arrested as the primary suspect in the murder of his wife Nicole, and Ronald Goldman. 95 million viewers watch Simpson's freeway chase in June.

-

DirecTV begins broadcasting an all-digital pay-TV system, by satellite, to the USA.

1995

Launch of Europe's first satellite for digital TV, Astra 1E, which marks the beginning of the age of digital television in Europe. It is also Astra's most powerful satellite to date.

-

Start of VIVA 2, VH-1, Hamburg 1, Nickelodeon, Super RTL and TM 3.

-

The first German teleshopping channel H.O.T. Hot Order Television goes on air.

1996

Astra 1F is launched (April 8, 1996), adding another 22 transponders (plus eight as back-up).

-

Germany's Premiere bundle of digital programme is offered on the satellite 1F. Thus, the next step in the operational trial for DVB begins (Digital Video Broadcasting).

-

EchoStar starts broadcasting its all-digital DBS pay-TV bouquet.

-

DF-1 starts a digital programme package.

-

Bertelsmann and RTL parent company CLT merge.

-

The tele-shopping broadcaster QVC goes on air.

1997

Astra 1G is launched (Dec 2, 1997) adding another 28 transponders to the fleet. Astra 5A/Sirius 2 is launched (November 12, 1997). Originally, the craft was known as Sirius 2 and the satellite transferred to SES Astra in April 2008 and placed at a new orbital position of 31.5 degrees East in order that Astra could exploit new markets in Central and Eastern Europe. Astra 5A was taken out of service on Jan 16, 2009.

-

Launch of a children's channel and Phoenix from ARD and ZDF.

1998

SES Astra adds another orbital position, 28.2 degrees East, to serve the UK and Irish markets. Astra 2A is launched (Aug 30, 1998) to serve this new position designed to cover the British Isles.

1999

Astra 1H is launched (June 18, 1999) adding a further 32 transponders to the fleet.

-

Launch of private news broadcaster N24.

-

CLT-UFA sells its Premiere shares in the Kirch Group, which consolidates Premiere and DF-1.

-

SES acquires 34% of the leading Asian satellite operating company AsiaSat, which supplies 53 countries with TV programmes.

-

CLT-UFA and the British Pearson plc amalgamated into the RTL Group, which is a shareholder of RTL, RTL II, Super RTL and VOX in Germany.

-

ProSieben and SAT.1 establish a broadcasting family (ProSieben, Kabel 1, N24 and Sat.1) under the umbrella of ProSiebenSat.1 Media AG.

2000

SES Astra acquires 50% of Scandinavian satellite operator Nordic Satellite Company (NSAB) in Stockholm.

-

Astra 2B is launched (Sept 14, 2000) to serve the UK and Irish markets.

-

Astra 2D is launched (Dec 19, 2000) again for the UK and Irish markets. The satellite is a Boeing Spinner type, bought on a rapid build contract from Boeing.

-

AOL and Time Warner merge to become the world's largest media player.

2001

For a total price of 4.3 billion dollars, SES takes over GE Americom, a US satellite operator and a subsidiary of General Electric Capital. For this, a new company, SES Global, is founded in Luxemburg. Through this merger, a worldwide leading satellite supplier emerges.

-

Astra 2C is launched (June 16, 2001) adding another 32 transponders to the fleet.

-

XM Satellite Radio begins service in the USA.

-

New York, the Twin Towers of the World Trade Center are attacked.

2002

Astra 1K is launched (Nov 25, 2002) but the upper stage (the Block DM-3 stage) of its giant Proton rocket fails to ignite correctly. It was de-orbited two weeks later on Dec 10. At the time Astra 1K was the world's largest commercial satellite and built by Alcatel Space. The plan for Astra 1K was for it to replace Astra 1B, and provide backup for 1A, 1C and 1D. The satellite also carried a wide compliment of Ka-band transponders and would have cleverly re-used some of these frequencies in multiple positions across Europe. Astra 1KR, its replacement, was successfully launched in 2006.

-

Astra 3A is launched (Mar 29, 2002) and destined for 23.5 degrees East, another new Astra orbital slot.

-

The Federal Cartel Office (FCO) prohibits the US media company Liberty Media from entering into the German TV cable network.

-

Sirius Satellite Radio starts transmissions in the USA.

-

The Kirch Group declares bankruptcy.

2003

Berlin-Brandenburg launches digital terrestrial television (DVB-T). It is the first time worldwide that analogue television via antenna is completely converted to digital transmittance in a metropolis.

-

The German TV group ProSiebenSat.1 is sold to US billionaire Haim Saban and investors after months of negotiations.

-

SES Global increases its stake in SES Sirius to 75%.

-

ARD and ZDF conclude a long-term contract with SES Global.

-

The Space Shuttle 'Columbia' breaks up on re-entry.

2004

SES Astra agrees with over 60 European broadcasting clients, hardware producers and further industry partners on technical specifications and a schedule for introducing HDTV services (High Definition Television) throughout Europe via the Astra satellite system. Europe's first HDTV channel is launched, Euro1080, supplied by Belgium's AlfaCam.

2005

SES Astra buys 100% of the DPC Digital Playout Centre from Premiere after the German Federal Cartel Office agrees.

-

Intelsat and PanAmSat merge.

2006

SES Astra increases its stake in ND SatCom to 100%.

-

Astra 1KR is launched (April 20, 2006) and as a replacement for the lost Astra 1K (see 2002). Astra 1KR is a powerful bird with 140 W of power. Astra 1A had just 45W available to it.

2007

SES Global is renamed SES and the company buys back the 19.5% stake held by General Electric.

-

Astra 1L is launched (May 4, 2007).

-

Astra 4A is launched (Nov 18, 2007) on a Proton rocket from Kazakhstan. Astra itself controls a 6-transponder portion of the overall 52 Ku-band transponder capacity on the satellite. The rest is held by sister company SES Sirius which operates its portion as Sirius 4. Astra's target is for this capacity is sub-Saharan Africa. The satellite sits at 5 degrees East.

2008

SES increases its ownership in SES Sirius to 90%. SES also develops another European orbital slot, at 31.5 degrees East.

-

Astra 1M is successfully launched (November 5, 2008).

Launch Posters

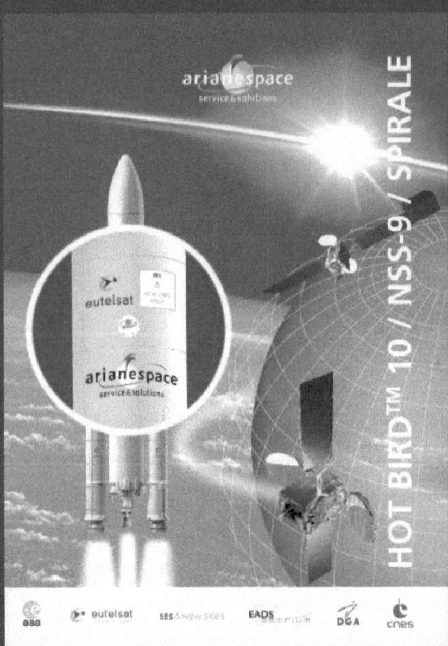

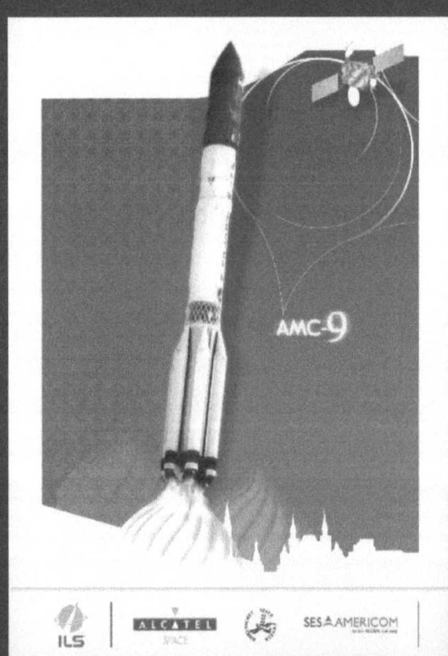

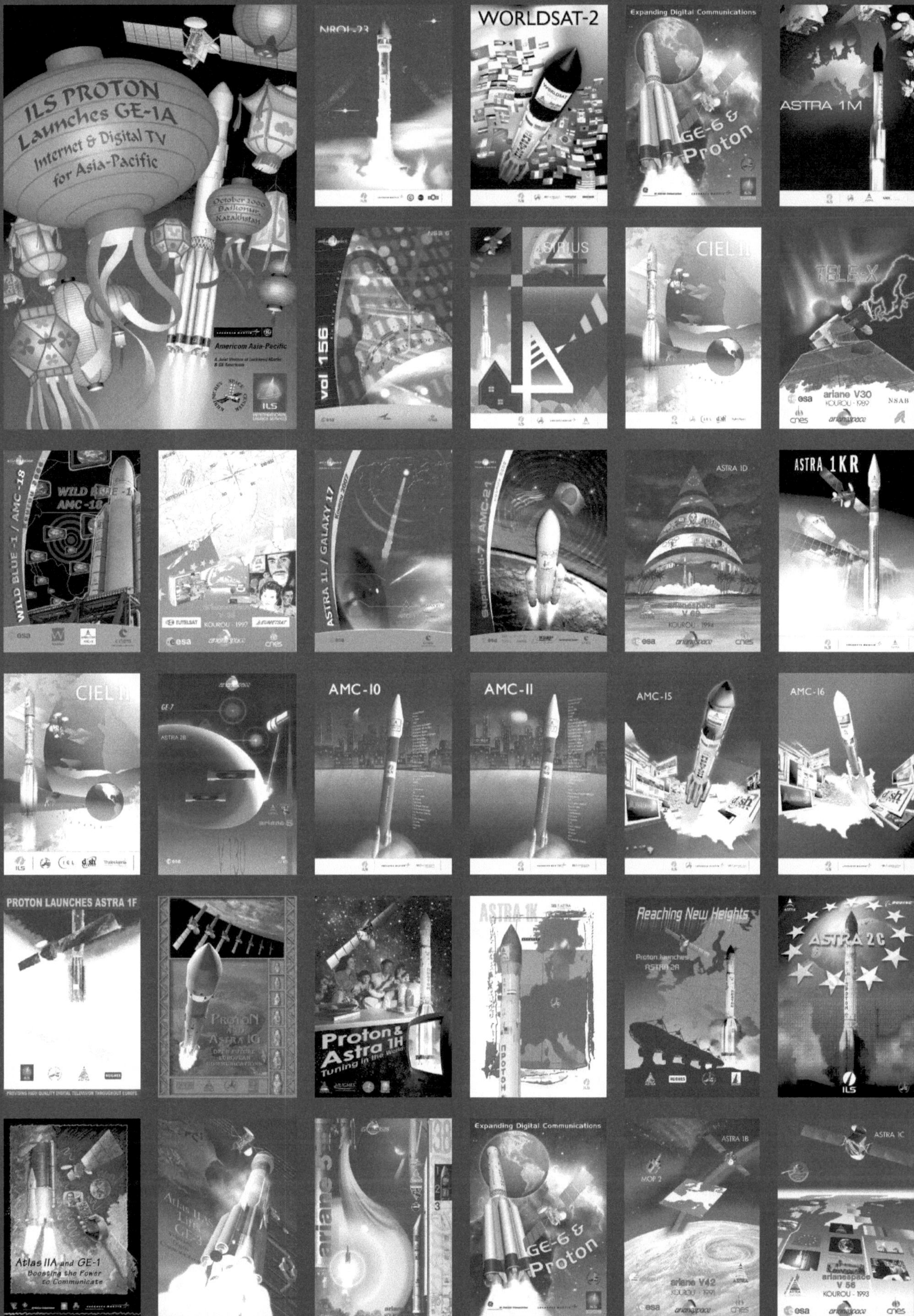
ILS PROTON
Launches GE-1A
Internet & Digital TV
for Asia-Pacific
October 2000
Baikonur, Kazakhstan
Americom Asia-Pacific
WORLDSAT-2
Expanding Digital Communications
GE-6 & Proton
ASTRA 1M
vol 156
SIRIUS
CIEL II
TELE-X
ariane V30
KOUROU - 1989
WILD BLUE-1 / AMC-18
KOUROU - 1997
ASTRA 1L / GALAXY 17
Superbird-7 / AMC-21
ASTRA 1D
KOUROU - 1994
ASTRA 1KR
CIEL II
GE-7
ASTRA 2B
ariane 5
AMC-10
AMC-11
AMC-15
AMC-16
PROTON LAUNCHES ASTRA 1F
Proton & Astra 1H
Tuning in the World
ASTRA 1K
Reaching New Heights
Proton launches ASTRA 2A
ASTRA 2C
Atlas IIA and GE-1
Boosting the Power to Communicate
ariane 5
Expanding Digital Communications
GE-6 & Proton
ASTRA 1B
MOP 2
ariane V42
ASTRA 1C
arianespace V 56
KOUROU - 1993

Acknowledgements

Contributors

London-based **Chris Forrester** is a well-known broadcasting journalist and industry consultant. He reports on all aspects of broadcasting with special emphasis on satellite, the business of television and emerging broadcast applications. This very much includes interactive multi-media and the growing importance of web-streamed and digitised content over all delivery platforms including satellite, cable and digital terrestrial TV as well as cellular and 3G mobile. Indeed, he has been investigating, researching and reporting on the so-called 'broadband explosion' for more than 25 years. He has been a freelance journalist since 1988. He was listening to short-wave radio when still in short trousers, and now says 'what is better than long-distance radio, but long-distance satellite reception!' He edits specialist newsletter Inside Satellite TV, as well as being Editorial Director at RapidTVNews.com. Chris is a member of the IBC Conference Committee.

Based near Cambridge in the UK, **Geoff Bains** has worked in technology journalism since 1983, investigating, reporting, and explaining the technological advances of the time, as they affect consumers – both in their homes and in the High Street. Author of several books on computing and electronics, Geoff began to follow the emerging satellite TV industry in 1986, and served as Editor of several specialist satellite TV publications, including a 13-year tenure at Europe's leading consumer magazine, What Satellite. Freelance since 2003, Geoff continues to inform and guide satellite TV viewers and enthusiasts both in print and online.

Julian Clover is a Media and Technology journalist based in Cambridge, UK. He has two decades of combined experience in online and printed media. Julian is an editor of Broadband TV News and New Television Insider. An accomplished conference moderator, Julian is a regular chairman at the annual IBC congress in Amsterdam, as well as Anga Cable in Cologne and the Broadband TV News Business Breakfasts. Television appearances include the BBC's technology magazine Click and the popular consumer programme Watchdog. He is a committee member of the Broadcasting Press Guild.

Dr. **Jörn Krieger** is based in Bavaria, Germany and has specialised in media journalism since 1990. He has contributed to a wide range of German and English-language trade publications, including Inside Satellite TV, Rapid TV News, Inside Digital TV and Interspace, in addition to providing consultancy and analysis. His emphasis is on satellite television, cable networks, digital television, pay-TV and new markets such as IPTV, HDTV and mobile TV.

In 2002, Jörn completed his doctorate thesis in communication studies and in the same year also published a handbook detailing the advantages of direct-to-home satellite broadcasting. In 2004, Jörn founded Medienbote, a weekly German-language media industry newsletter offering exclusive news, interviews and analysis which has quickly developed into an indispensable source of information for Europe's top media executives.

Serge Siritzky studied at ENA, one of France's Grandes Écoles. Living in Paris, he has dual French and American nationality, and has a long career in the cinema and television industries. He is currently Editorial Director of Ecran Total, the highly-regarded French magazine devoted to the audio-visual business.

Every book is a collaboration, and 'High Above' represents a great deal more than the usual collective effort of a half-dozen people. This has been a massive undertaking, and it simply would not have been possible without the unstinting help of SES Astra's team of experts. To Romain Bausch and Ferdinand Kayser who gave it their enthusiastic support. To Markus Payer I owe an immense debt of gratitude. His colleagues James Collins, Jean-Paul Hoffmann, Yves Feltes, Nick Stubbs, Mike Chandler, Paul Freeman, Fred Dietzel, Antonio Lopa, David Netterville, Kirstin Steffen and Katharina Staszkow, and many others at Betzdorf all gave the project their enthusiastic and generous help.

My thanks also go to Mario de Lépine at Arianespace and Matthieu Duvelleroy at EADS-Astrium.

'High Above' would have stayed firmly on the ground had it not been for Chelsea Spieker and Ralf Beke-Bramkamp at 'plan B communication' in Berlin. Our designers coped heroically with the 101 changes needed on such a complex task, and in particular Harald Lipken and Birte Buschmeier at Triad, Berlin, as well as Sebastian Stottele, who were wonderful partners on the project.

Finally, and not least, to my wife Marianne, who proof-read, encouraged me and tolerated my late nights and very early mornings on the keyboard as the book came together.

Our historical trawl for documents also saw us sourcing archive material from AlfaCam, Arqiva, BBC, BSkyB, Canal+, EMI, RTL, Globecast, Panasonic, Philips, LG, Technisat and a number of web-sites, all of which generously permitted us to use their material. 'High Above' gratefully acknowledges the use of TV images from the following sources:

Meldrum Home Page
(www.meldrum.co.uk)

Chuck Pharis, of Pharis Video
(www.pharis-video.com)

TV History
(www.tvhistory.tv)

Early Technology Collection
(TVIK)

Old Radio
(www.oldradio.com)

Mike Bennet's TV Museum
(www.oldtechnology.com)

Paul Schatzkin
(www.farnovision.com)

photo credits:
Frauenhofer HHi Time Lab
Getty Images
Ullstein Bild
Arianespace
EADS-Astrium
SES Astra
Esa (ESA Multimedia)
Thales Alenia Space

Richmond, West London
March 2010

The chapters open with the following photos:

Opening ceremony of Olympic Games 2008, Beijing, China **(Chapter 1)**, Astra's dish park in Betzdorf, Luxembourg **(Chapter 2)**, Astronaut conducts tests in space at 130 nautical miles distance from the Earth **(Chapter 3)**, Rocket launch in Kourou, French Guiana **(Chapter 4)**, Olympic Games 2010 in Vancouver **(Chapter 5)**, Satellite dishes at an apartment block in Berlin **(Chapter 6)**, "Trabi" in front of the former inner-German border **(Chapter 7)**, SAT.1 balloon above Potsdamer Platz in Berlin **(Chapter 8)**, Penelope Cruz at Cannes Film Festival in 2009 **(Chapter 9)**, Rocket liftoff at Kourou, French Guiana **(Chapter 10)**, Astra's dish park in Betzdorf, Luxembourg **(Chapter 11)**, Integration of the satellite into the top of the rocket **(Chapter 12)**, Satellite in test chamber in Astrium's plant in Toulouse **(Chapter 13)**, Reporter in front of the Bird Cage at the Olympics in Beijing, China, 2008 **(Chapter 14)**, Visitors to IFA trade show in Berlin **(Chapter 15)**, A view from above Tokyo **(Chapter 16)**, Visitors to a multiplex cinema **(Chapter 17)**, In the belly of Astra's control rooms **(Chapter 18)**.

Bibliography

Our suggested texts are far from complete but readers with a deeper thirst for knowledge might find some of our suggestions useful.

Achilléas, Philippe (1997)
La télévision par satellite: aspects juridiques internationaux.
LGDJ / Montchrestien.

Ahrens, Wilfried (1998)
Astra - Fernsehen ohne Grenzen. Eine Chronik.
Econ Verlag.

Alberty, Michael, Siegfried Börner and Gerhard Ohm (1984)
Entwicklung eines 20-GHz-Modulators für die 20-, 30-GHz-Nutzlast eines Satelliten für Ausbreitungs- und Übertragungsexperimente einschließlich der erforderlichen Umsetzer für die Testeinrichtung.
Eggenstein-Leopoldshafen, Fachinformationszentrum Energie, Physik, Mathematik GmbH Karlsruhe.

Altendorfer, Otto (2004)
Das Mediensystem der Bundesrepublik Deutschland. Band 2.
VS Verlag.

Arnett, Peter (1994)
Live from the Battlefield.
Simon and Schuster.

C. Forrester (ed.), *High Above: The Untold story of Astra, Europe's leading satellite company*,
DOI 10.1007/978-3-642-12009-1, © Springer-Verlag Berlin Heidelberg 2011

Auletta, Ken (2004)
Media Man.
Atlas Books.

Auletta, Ken (2003)
Back Story: Inside the Business of News.
The Penguin Press.

Banks, Jack (1996)
Monopoly Television.
Westview.

Banwait, Narinder (1999)
Television in Europe to 2008.
Zenith Media.

Barnett, Steven (1994)
The Battle for the BBC.
Aurum Press.

Baumann, Ingo (2005)
Das internationale Recht der Satellitenkommunikation.
Peter Lang.

Baumgart, Michael and Hans Dodel (1990)
Satellitensysteme für Kommunikation, Fernsehen und Rundfunk: Theorie und Technologie.
Hüthig.

Belfield, Richard, et al (1991)
Murdoch: The Great Escape.
Warner Books.

Berger, Viktor J.F. (2008)
Der deutsche Fernsehmarkt.
Paderborn. Wilhelm Fink.

Bibb, Porter (1994)
It Ain't as Easy as it Looks: Ted Turner's Story
Virgin Books.

Biederman, Donald E. et al. (2007)
Law and Business of the Entertainment Industries. 5th Edition.
CT: Praeger Publishers.

Blumenthal, Howard (1991)
This Business of Television.
Billboard.

Blumenthal, Howard J. and Oliver R. Goodenough (2006)
This Business of Television: The Standard Guide to the Television Industry.
Billboard Books.

Bower, Tom (1995)
Maxwell: the Final Verdict.
Harper Collins.

Brice, Richard (2001)
Newnes Guide to Digital Television.
Newnes.

Brinkley, Joel (1997)
Defining Vision: the Battle for the Future of Television.
Harcourt Brace.

Brown, Maggie (2007)
A Licence to be Different.
British Film Institute.

Boyd, Andrew (1988)
Broadcast Journalism.
Heinemann.

Busch, Gernot (2009)
Die Einführung des Satellitendirektempfangs in Deutschland über Astra: Ein Ergebnis der Push-Pull-Strategie bei der Diffusion von Empfangsgeräten im Konsumenten-Markt.
VISTAS Verlag.

CasaBianca, Lou (1992)
The New TV: A Comprehensive Survey of High Definition Television.
CT: Meckler.

Clarke, Arthur C. (1965)
Voices from the Sky.
Harper & Row.

Clarke, Arthur C. (1992)
How the World was One.
Victor Gollancz.

Clarke, Arthur C. (1993)
By Space Possessed.
Victor Gollancz.

Carter, Bill (1995)
The Late Shift.
Hyperion Books.

Chartrand, Mark R. (2004)
Satellite Communications for the Nonspecialist.
The International Society for Optical Engineering.

Chayes, Abram (1973)
Satellite Broadcasting.
New York/ Oxford: Oxford University Press.

Chayes, Abram, Monroe Edwin Price and International Broadcast Institute (1975)
Direct Broadcasting from Satellites: Policies and Problems.
West Publishing Company.

Chenoweth, Neil (2001)
Virtual Murdoch.
Secker and Warburg.

Chippindale, Peter (1991)
Dished!: The Rise and Fall of BSB.
Simon and Schuster.

Coleman, James A. and Rollet, Brigitte (1997)
Television in Europe.
Exeter: Intellect Books.

Collins, Richard (1992)
Satellite Television in Western Europe.
John Libby.

Collins, Richard (1993)
Broadcasting and Audio-Visual Policy in the European Single Market.
Luton: University of Luton Press.

Collins, Richard (1998)
From Satellite to Single Market: New Communication Technology and European Public Service Television.
Routledge.

Cockerell, Michael (1988)
Live from Number 10.
Faber and Faber.

Colzi, Enrico (1999)
Digital Signal Processing Techniques for Personal and Broadcasting Satellite.
Communication Systems. ESA Publ. Div. c/o ESTEC.

Cones, John (1992)
Film Finance and Distribution.
Silman-James Press.

Crisell, Andrew (1997)
An Introductory History of British Broadcasting.
Routledge.

Curtis, Anthony R. (1994)
Space satellite handbook. 3rd Edition.
Gulf Publishing.

Davies, J.K. (1992)
Space Exploration.
W&R Chambers.

Deliyanni, Elsa (1993)
Le droit de représentation des auteurs face à la télévision transfrontalière par satellite et par cable.
Libraire Générale de Droit et de Jurisprudence.

d'Haenens, Leen and Frieda Saeys (2007)
Western Broadcast Models: Structure, Conduct and Performance.
de Gruyter.

Dittel, Volker, Friedrich Manz and Jean Pütz (1989)
Alles über Fernsehen, Video, Satellit.
VGS Verlag.

Donovan, Paul (1997)
All Our Todays.
Jonathan Cape.

Doran, Jamie and Bizony, Piers (1998)
Starman. The Truth Behind The Legend Of Yuri Gagarin.
Bloomsbury Publishing.

Dover, Bruce (2008)
Rupert's Adventures in China.
Mainstream.

Doyle, Marc (1992)
The Future of Television: A Global Overview of Programming, Advertising, Technology, and Growth.
McGraw-Hill Contemporary.

Dowmunt, Tony (1993)
Channels of Resistance: Global Television and Local Empowerment.
BFI Publishing in association with Channel Four Television.

Dupagne, Michel and Peter Benjamin Seel (1998)
High-Definition Television: A Global Perspective.
Iowa State University Press.

Dyke, Greg (2004)
Inside Story.
Harper Collins.

Dyson, Kenneth and Peter Humphreys (1990)
The Political Economy of Communications. International and European Dimensions.
Routledge.

Dyson, Kenneth and Peter Humphreys et al. (1988)
Broadcasting and New Media Policies in Western Europe.
Routledge.

Elbert, Bruce R. (2004)
The Satellite Communication Applications Handbook.
Artech House, Inc.

Euromedia Research Group, Denis McQuail and Karen Siune (1999)
Media Policy: Convergence, Concentration & Commerce.
SAGE Publications Ltd.

Evans, Barry G. (1999)
Satellite Communication Systems.
Institution of Electrical Engineers.

Falk, Quentin (2005)
Television's Strangest Moments.
Robson Books.

Fisher, David I. (1990)
Prior Consent to International Direct Satellite Broadcasting.
Dordrecht/Kluwer Academic Publishers.

Fischer, Walter (2008)
Digital Video and Audio Broadcasting Technology: A Practical Engineering Guide.
Springer.

Flath, Herbert (2007)
Fernsehen heute und in der Zukunft: Digitalfernsehen, Kabel- und Satellitengesellschaften, IPTV und Handy-TV.
Norderstedt: Books on Demand GmbH.

Forrester, Chris (2000)
The Business of Digital Television.
Focal Press.

Friedmann, Annette (2000)
Das erste digitale Fernsehen in Deutschland: Voraussetzungen und Programmpotentiale des digitalen Pay-TVs anhand der Markterfahrungen von DF 1.
Verlag für Wirtschaftskommunikation.

Ford, Bianca and James Ford (1993)
Television and Sponsorship.
Butterworth-Heinemann.

Gilder, George (1990)
Life after Television.
Norton Books.

Gitlin, Todd (1983)
Inside Prime Time.
Routledge.

Goldberg, Robert (1995)
Citizen Turner.
Harcourt Brace.

Goodwin, Peter (1998)
Television under the Tories.
The British Film Institute.

Griffiths, Alan (2003)
Digital Television Strategies.
Paul Grave Macmillan.

Goleniewski, Lillian (2003)
Telecommunications Essentials: The Complete Global Source for Communications Fundamentals, Data Networking and the Internet, and Next-Generation Networks.
Pearson Education, Inc.

Gomery, Douglas and Luke Hockley (2008)
Television Industries.
British Film Institute.

Gunther, Marc (1994)
The House that Roone Built.
Little, Brown.

Haagen, Bernd-Ulrich (1986)
Satellitenfernsehen: Rechtsprobleme einer neuen Technologie.
Peter Lang.

Hack, Richard (2003)
Clash of the Titans.
The New Millennium Press.

Hammond, Ray (1996)
Digital Business.
Hodder and Stoughton.

Harcourt, Alison (2005)
The European Union and the Regulation of Media Markets.
Manchester University Press.

Harrison, Martin (1985)
TV News: Whose Bias?
Policy Journals.

Hart, Jeffrey A. (2007)
Technology, Television and Competition. The Politics of Digital TV.
Cambridge University Press.

Hasebrink, Uwe (2001)
Fernsehen in neuen Medienumgebungen. Befunde und Prognosen zur Zukunft der Fernsehnutzung.
VISTAS Verlag.

Hasebrink, Uwe, Hermann-Dieter Schröder and Birgit Stark (2008)
Elektronische Programmführung im digitalen Fernsehen. Nutzerstudie und Marktanalyse.
VISTAS Verlag.

Hege, Hans (1998)
Digitales Fernsehen in USA und Europa.
Wissenschaftlicher Verlag Trier.

Henle, Victor (1998)
Fernsehen in Europa: Strukturen, Programme und Hintergründe.
Kopaed.

Herrmanns, Linda (2007)
Fernsehen ohne Grenzen: der deutsche TV-Markt zwischen Qualität und Quote.
Tectum Verlag.

Hetherington, Alastair (1985)
News, Newspapers and Television.
Macmillan.

Higgins, Jonathan (2007)
Satellite Newsgathering.
Burlington, MA/ Oxford: Focal Press.

Holland, Patricia (1997)
The Television Handbook.
London/ New York: Routledge.

Holtmann, Klaus (1999)
Programmplanung im werbefinanzierten Fernsehen.
Josef Eul Verlag GmbH.

Holznagel, Bernd, Dieter Dörr and Doris Hildebrand (2008)
Elektronische Medien: Entwicklung und Regulierungsbedarf.
Vahlen.

Horrie, Chris and Clarke, Steve (1994)
Fuzzy Monsters: Fear and Loathing at the BBC.
Heinemann.

Horrie, Chris (2000)
Citizen Greg.
Simon and Schuster.

Horrie, Chris (1999)
Live TV: The Uncut Story of Tabloid Television.
Simon and Schuster.

Horsman, Mathew (1997)
Sky High: the Inside Story of BSkyB.
Orion Business.

Hübner, Heinz (1983)
Satellitenfernsehen und deutsches Rundfunksystem.
Munich: C.H. Beck.

Hüffel, Clemens and Anneliese Rohrer (2007)
Hat öffentlich-rechtliches Fernsehen Zukunft?
Holzhausen Verlag.

Humphreys, Peter (1994)
Media and Media Policy in Germany. The Press and Broadcasting since 1945. 2nd Edition.
Berg Publishers.

Humphreys, Peter J. (1996)
Mass Media and Media Policy in Western Europe.
Manchester University Press.

Humphreys, Peter and Seamus Simpson (2005)
Globalisation, Convergence and European Telecommunications Regulation.
Edward Elgar Publishing.

Isaacs, Jeremy (2006)
Look Me in the Eye.
Little, Brown.

Johnson, J O and Martine Orange (2003)
The Man Who Tried to Buy the World.
Viking.

Johnston, Carla Brooks (1995)
Winning the Global TV News Game.
Focal Press.

Kadish, Jules E. and Thomas W.R. East (2000)
Satellite Communications Fundamentals.
Artech House.

Kaumanns, Ralf, Veit Siegenheim and Insa Sjurts (2008)
Auslaufmodell Fernsehen? Perspektiven des TV in der digitalen Medienwelt.
GWV Fachverlage GmbH.

Karstens, Eric and Jörg Schütte (2005)
Praxishandbuch Fernsehen: Wie TV-Sender arbeiten.
VS Verlag für Sozialwissenschaften.

Kelly, Mary, Gianpietro Mazzoleni and Denis McQuail (2004)
The Media in Europe: The Euromedia Handbook.
SAGE Publications Ltd.

Kim, Myung Joong (1989)
Satellitenfernsehen in Europa: unter besonderer Berücksichtigung des Programmangebots, der Modelle der Zusammenarbeit und des Programmauftrags der übernationalen, nicht-privatwirtschaftlich organisierten Satellitenprogrammanbieter.
Peter Lang.

King, Larry (1995)
The Best of Larry King Live.
Turner.

Kipps, Charles (1989)
Out of Focus.
Mackays.

Klein, Alec (2003)
Stealing Time.
Simon and Schuster.

Kleine-Erfkamp, Stefan (1996)
Szenarioanalyse für die Verbreitung von Fernsehprogrammen über Terrestrik, BK-Netze und Satellit in der BRD bis ins Jahr 2000.
Löw & Vorderwülbecke.

Kluger, Jeffrey (1995)
The Apollo Adventure.
Boxtree.

Krieg, Bernhard (1993)
Satellitenfernsehen: wenig Theorie - viel Praxis. 4th Edition.
Elektor-Verlag.

Krieger, Jörn (2002)
Fernsehen, Radio, Internet aus dem Orbit. Die Vorteile des Satellitenempfangs.
Vth Verlag.

Kuhn, Raymond (1985)
Broadcasting and Politics in Western Europe.
Frank Cass & Co. Ltd.

Kurtz, Howard (1996)
Hot Air.
Times Books.

Kurtz, Howard (2007)
Reality Show: The Last Great Television News War.
Simon and Schuster.

Langheinrich, Thomas, Hans Hege and Arbeitsgemeinschaft der Landesmedienanstalten (ALM) und Gemeinsame Stelle digitaler Zugang (GSDZ) (2008)
Digitalisierungsbericht 2008. Die Frage nach dem digitalen Mehrwert. Neue Inhalte und ihre Finanzierung.
Vistas Verlag.

Lau, Sebastian (2007)
Standardisierung und Wettbewerbsstrategien im interaktiven Fernsehen.
Cuvillier Verlag.

Leapman, Michael (1986)
The Last Days of the Beeb.
Allen and Unwin.

Le Gouëff, Stéphan (2001)
Satellite Regulation in Europe: Legal Texts and Materials.
Kluwer Law International.

Lenhardt, Helmut (1972)
Die Zukunft von Rundfunk und Fernsehen. In der Auseinandersetzung mit den neuen elektronischen Medien.
Molden.

Leonhard, Joachim-Felix (1999)
Medienwissenschaft: ein Handbuch zur Entwicklung der Medien und Kommunikationsformen. Teilbd. 1.
de Gruyter.

Leopoldt, Swaantje (2002)
Navigatoren: Zugangsregulierung bei elektronischen Programmführern im digitalen Fernsehen.
Nomos Verlagsgesellschaft.

Levy, David A. L. (1999)
Europe's Digital Revolution: Broadcasting Regulation, the EU and the Nation State.
Routledge.

Long, Mark (1998)
The World of Satellite Television.
Mark Long Enterpries.

Long, Mark E. (1986)
The Ku-band Satellite Handbook.
Indianapolis, IN: H.W. Sams.

Long, Mark E. (1999)
The Digital Satellite TV Handbook.
Woburn, MA: Butterworth-Heinemann.

Luther, Sara Fletcher (1988)
The United States and the Direct Broadcast Satellite: The Politics of International Broadcasting in Space.
Oxford University Press.

Lyons, Eugene (1966)
David Sarnoff.
Harper and Row.

Mahler, Gerhard (2005)
Die Grundlagen der Fernsehtechnik: Systemtheorie und Technik der Bildübertragung.
Springer.

Maral, Gerard and Michel Bousquet (2002)
Satellite Communications Systems: Systems Techniques and Technology. 4th Edition.
John Wiley & Sons.

Martin, Donald H. (2000)
Communication Satellites.
Los Angeles: Aerospace Press.

Matzen, Christiane (2000)
Internationales Handbuch für Hörfunk und Fernsehen, 2000/2001.
Nomos Verlagsgesellschaft.

McAleer, Neil (1992)
Odyssey. The Authorised Biography of Arthur C. Clarke.
Victor Gollancz.

McCormac, John (1996)
European Scrambling Systems.
Waterford University Press.

Meyn, Hermann (1996)
Massenmedien in der Bundesrepublik Deutschland.
Colloquium.

Miall, Leonard (1994)
Inside the BBC.
Weidenfield and Nicolson.

Monkenbusch, Helmut (1994)
Medien, Macht und Märkte.
Reinbek: Rowohlt.

Negrine, Ralph M. (1988)
Satellite Broadcasting. The Politics and Implications of the New Media.
Routledge.

Neil, Andrew (1996)
Full Disclosure.
Macmillan.

Newcomb, Horace (2004)
Encyclopedia of Television.
Fitzroy Dearborn.

NHK Science and Technical Research Laboratories (1993)
High Definition Television: Hi-Vision Technology.
Kluwer Academic Publishers.

Noam, Eli (1991)
Television in Europe.
Oxford/ New York: Oxford University Press.

Ohmori, Shingo, Hiromitsu Wakana and Seiichiro Kawase (1997)
Mobile Satellite Communications.
Artech House.

Ott, Alexandra (2008)
Der deutsche und japanische Fernsehmarkt: Vergleichende Analyse und prognostizierte Entwicklung.
VDM Verlag.

Page, Bruce (2003)
The Murdoch Archipelago.
Simon and Schuster.

Parks, Lisa (2005)
Cultures in Orbit: Satellites and the Televisual.
Duke University Press.

Parsons, Patrick R. and Robert M. Frieden (1997)
The Cable and Satellite Television Industries.
Allyn and Bacon.

Pascall, Stephan C. and David J. Withers (2002)
Commercial Satellite Communication. 2nd Edition.
Focal Press.

Pattan, Bruno (1993)
Satellite systems: Principles and Technologies.
Van Nostrand Reinhold.

Paukens, Hans and Andreas Schümchen (2000):
Digitales Fernsehen in Deutschland: Explorative Studie zur Entwicklung digitaler Pay-TV-Angebote.
R. Fischer.

Pelton, Joseph N. (2006)
The Basics of Satellite Communications. International Engineering.
Consortium.

Pelton, Joseph N. (1995)
Wireless and Satellite Telecommunications.
Prentice-Hall.

Pelton, Joseph N., Peter P. Marshall and Robert J. Oslund (2004)
Communications Satellites: Global Change Agents.
Lawrence Erlbaum Associates.

Picard, Robert G. and Allan Brown (2005)
Digital Terrestrial Television in Europe.
Lawrence Erlbaum Associates.

Pichler, Marie Helen (1987)
Copyright Problems of Satellite and Cable Television in Europe.
Graham & Trotman; M. Nijhoff.

Pierce, John Robinson (1968)
The Beginnings of Satellite Communications.
San Francisco Press Inc.

Platho, Rolf (1999)
Fernsehen und Hörfunk transparent: Recht, Wirtschaft, Programm, Technik.
R. Fischer.

Plog, Philipp (2005)
Zugang zum digitalen Fernsehen in Frankreich.
Nomos Verlagsgesellschaft.

Porter, Henry (1984)
Lies, Damned Lies and Some Exclusives.
Chatto and Windus.

Powell, Jon T. (1985)
International Broadcasting by Satellite: Issues of Regulation, Barriers to Communication.
Quorum Books.

Pritchard, Wilbur, Henri Suyderhoud and Robert A. Nelson (1993)
Satellite Communication Systems Engineering.
Prentice-Hall.

Prütting, Hanns (2000)
Marktmacht und Konzentrationskontrolle auf dem Fernsehmarkt.
C.H. Beck.

Queeney, Kathryn M. (1978)
Direct Broadcast Satellites and the United Nations.
Sijthoff & Noordhoff International Publishers B.V.

Redstone, Sumner (2001)
A Passion to Win.
Simon and Schuster.

Reimers, Ulrich (2008)
DVB - Digitale Fernsehtechnik: Datenkompression und Übertragung.
Springer.

Rennenberg, Christiane (2007)
Digitalisierung der Unterhaltungselektronik. Das Fernsehen der Zukunft.
VDM Verlag.

Reinshagen, Urs M. (1971)
Satellitensendungen und Urheberrecht. Ein Beitrag zur rechtlichen Problematik von Weltraumübertragungen.
Zürich Schulthess Polygraphischer Verlag.

Rzeszewski, Theodore S. (1995)
Digital Video: Concepts and Applications Across Industries.
Institute of Electrical and Electronics Engineers.

Rha, Nak Gyun (1998)
Die grenzüberschreitende Programmübertragung des Satellitenrundfunks und dessen urheberrechtliche Probleme.
Freie Universität Berlin.

Richardson, Kay and Ulrike H. Mainhof (1999)
Worlds in Common? Television Discourses in a Changing Europe.
Routledge.

Riegler, Thomas (2006)
Grundlagen des Satellitenempfangs: Ratgeber für Einsteiger und Fortgeschrittene.
Vth Verlag.

Riegler, Thomas (2006)
Programmvielfalt via Satellit: Wie man Fernsehen und Radio aus aller Welt empfängt.
Vth Verlag.

Roddy, Dennis (2006)
Satellite Communications. 4th Edition.
McGraw Hill.

Rogers, Everett M. (1986)
Communication Technology: The New Media in Society.
Free Press.

Rozell, Mark J. (2003)
Media Power, Media Politics.
Rowman & Littlefield Publichers, Inc.

Roßnagel, Alexander, Werner Sosalla and Thomas Kleist (2004)
Der Zugang zur digitalen Satellitenverbreitung: Digitalisierung und Fernsehen ohne Grenzen. Gutachten im Auftrag der Gemeinsamen Stelle Digitaler Zugang der Landesmedienanstalten.
VISTAS Verlag.

Samuel, Lawrence R. (2001)
Post War Television Advertising.
Texas.

Schonfeld, Reese (2001)
Me and Ted against the World.
Harper Collins.

Schaefer, Todd M. and Thomas A. Birkland (2007)
Encyclopedia of Media and Politics.
CQ Press.

Schenk, Michael, Thomas Döbler and Birgit Stark (2002)
Marktchancen des digitalen Fernsehens: Akzeptanz und Nutzung von Pay-TV und neuen Diensten.
Westdeutscher Verlag.

Schmid, Andreas and Johannes Ebert (2002)
50 Jahre deutsches Fernsehen.
Deutsche Post.

Schmidbauer, Michael (1983)
Satellitenfernsehen für die Bundesrepublik Deutschland: Bedingungen und Möglichkeiten des Direktfernsehens via Satellit.
Spiess Volker GmbH.

Schnitzler, Daniel J. (2008)
Digitaler Fernsehmarkt: Chancen und Potentiale für Spartensender.
Diplomica Verlag GmbH.

Schulz, Dieter (2007)
Das inoffizielle Sat-Receiver-Buch.
Franzis Verlag GmbH.

Shephard, Alan (1994)
Moon Shots.
Turner.

Siebel, Wolf and Michael Schmitz (2008)
Sender und Frequenzen 2009: Jahrbuch für weltweiten Rundfunkempfang. Lang-, Mittel-, Kurzwelle. Satellit. Internet. Inklusive Empfängertests und Programmtipps.
Vth Verlag.

Siebenhaar, Hans-Peter (1994)
Europa als audiovisueller Raum: Ordnungspolitik des grenzüberschreitenden Fernsehens in Westeuropa.
Leske + Budrich.

Signitzer, Benno (1976)
Regulation of direct broadcasting from satellites: the U.N. involvement.
Praeger.

Simpson, Wes and Howard Greenfield (2007)
IPTV and Internet Video.
Focal Press.

Siune, Karen, Wolfgang Truetzschler, Euromedia Research Group (1992)
Dynamics of Media Politics: Broadcast and Electronic Media in Western Europe.
SAGE Publications Ltd.

Slamin, Brendan (2008)
High Definition TV: the Essential Guide.
The Digital TV Group.

Smolders, Peter (1973)
Soviets in Space.
Butterworth Press.

Snoddy, Raymond (1992)
The Good, the Bad and the Unacceptable.
Faber and Faber.

Squire, Jason (1992)
The Movie Business Book.
Simon and Schuster.

Stark, Birgit (2006)
Fernsehen in digitalen Medienumgebungen: Eine empirische Analyse des Zuschauerverhaltens.
R. Fischer.

Stashower, Daniel (2002)
The Boy Genius and the Mogul.
Broadway Books.

Steinmetz, Rüdiger (1996)
Freies Fernsehen: das erste privat-kommerzielle Fernsehprogramm in Deutschland.
UVK Medien.

Stephenson, D.J. (1990)
Guide to Satellite Televison.
Newnes.

Stewart, M. LeSueur (1989)
To See the World: The Global Dimension in International Direct Television Broadcasting by Satellite.
Kluwer Law International.

Straubhaar, Joseph D. (2007)
World Television: From Global to Local.
Sage.

Trebbe, Joachim (2004)
Fernsehen in Deutschland 2003-2004: Programmstrukturen, Programminhalte, Programmentwicklungen: Forschungsbericht.
VISTAS Verlag.

Van Tassel, Joan (2001)
Digital TV over Broadband.
Focal Press.

Wallis, Roger (1990)
The Known World of Broadcast News.
Routledge.

Ward, David (2002)
The European Union Democratic Deficit and the Public Sphere. An Evaluation of EU Media Policy.
OS Press.

Wacker, Katharina (2007)
Wettbewerb und Regulierung auf dem deutschen Fernsehmarkt: Deregulierungsbedarf und Umsetzungsbedingungen.
Lucius & Lucius.

Whalen, David J. (2002)
The Origins of Communications Satellite Technology (1945-1965).
Smithsonian Press.

Whittemore, Hank (1990)
CNN: The Inside Story.
Little, Brown.

Winans, Christopher (1995)
The King of Cash (the Story of Laurence Tisch).
Wiley.

Winkler, Klaus and Arbeitsgemeinschaft für Kommunikationsforschung e. V. (1984)
ommerzielles Fernsehen in der Medienkonkurrenz. Band 2.
Spiess.

Wilson, Roger (1994)
Local Television, Finding a Voice.
Dragonflair.

Wolff, Michael (2008)
The Man Who Owns the News.
Bodley Head.

Wood, James (1992)
History of International Broadcasting.
Peter Peregrinus Ltd.

Wyatt, Will (2003)
The Fun Factory: Life in the BBC.
Aurum Press.

Zeller, Thomas (2002)
Size Matters: Die Fusionen in der amerikanischen Medienindustrie und ihre Auswirkungen auf den deutschen Fernsehmarkt.
Verl. für Interkulturelle Kommunikation.

Zervos, Frank (2003)
Digitales Fernsehen in Deutschland: medienpolitische und medienwirtschaftliche Herausforderungen des zukünftigen Fernsehens.
Westdeutscher Verlag.

Ziemer, Albrecht (1997)
Digitales Fernsehen: Eine neue Dimension der Medienvielfalt. Ausgabe 2.
Hüthig.

Zimmer, Jochen (1993)
Europas Fernsehen im Wandel: Probleme einer Europäisierung von Ordnungspolitik und Programmen.
Peter Lang.

Published by
Broadgate Publications
37 The Towers
Lower Mortlake Rd
Richmond, UK
TW9 2JR